# Remapping Urban Heat Island Atlases in Regenerative Cities

Hisham Abusaada
*Housing and Building National Research Center, Egypt*

Abeer Elshater
*Ain Shams University, Egypt*

Marwa Khalifa
*Ain Shams University, Egypt*

A volume in the Advances in Civil and Industrial Engineering (ACIE) Book Series

IGI Global
PUBLISHER of TIMELY KNOWLEDGE

Published in the United States of America by
IGI Global
Engineering Science Reference (an imprint of IGI Global)
701 E. Chocolate Avenue
Hershey PA, USA 17033
Tel: 717-533-8845
Fax: 717-533-8661
E-mail: cust@igi-global.com
Web site: http://www.igi-global.com



Library of Congress Cataloging-in-Publication Data

Names: Abusaada, Hisham, DATE- editor. | Elshater, Abeer, DATE- editor. | Khalifa, Marwa, DATE- editor.
Title: Remapping urban heat islands atlases in regenerative cities / Hisham Abusaada, Abeer Elshater, and Marwa Khalifa, editors.
Description: Hershey, PA : Engineering Science Reference, [2022] | Includes bibliographical references and index. | Summary: "This book focuses on five SDGs: poverty (Goal 1), public health and well-being (Goal 3), decent work and economic growth (Goal 8), sustainable cities and societies (Goal 11), and climate action (Goal 13)that are associated with the increasing Urban Heat Island phenomenon that accompanies rapid urbanization, which has changed the way of life of many countries worldwide"-- Provided by publisher.
Identifiers: LCCN 2022001264 (print) | LCCN 2022001265 (ebook) | ISBN 9781668424629 (hardcover) | ISBN 9781668424636 (paperback) | ISBN 9781668424643 (ebook)
Subjects: LCSH: Urban heat island. | Urban climatology. | Urban ecology (Sociology)
Classification: LCC QC981.7.U7 R46 2022 (print) | LCC QC981.7.U7 (ebook) | DDC 551.5/25091732--dc23/eng20220317
LC record available at https://lccn.loc.gov/2022001264
LC ebook record available at https://lccn.loc.gov/2022001265

This book is published in the IGI Global book series Advances in Civil and Industrial Engineering (ACIE) (ISSN: 2326-6139; eISSN: 2326-6155)

British Cataloguing in Publication Data
A Cataloguing in Publication record for this book is available from the British Library.

All work contributed to this book is new, previously-unpublished material.
The views expressed in this book are those of the authors, but not necessarily of the publisher.

For electronic access to this publication, please contact: eresources@igi-global.com.

# Advances in Civil and Industrial Engineering (ACIE) Book Series

ISSN:2326-6139
EISSN:2326-6155

Editor-in-Chief: Ioan Constantin Dima University Valahia of Târgovişte, Romania

## MISSION

Private and public sector infrastructures begin to age, or require change in the face of developing technologies, the fields of civil and industrial engineering have become increasingly important as a method to mitigate and manage these changes. As governments and the public at large begin to grapple with climate change and growing populations, civil engineering has become more interdisciplinary and the need for publications that discuss the rapid changes and advancements in the field have become more in-demand. Additionally, private corporations and companies are facing similar changes and challenges, with the pressure for new and innovative methods being placed on those involved in industrial engineering.

The **Advances in Civil and Industrial Engineering (ACIE) Book Series** aims to present research and methodology that will provide solutions and discussions to meet such needs. The latest methodologies, applications, tools, and analysis will be published through the books included in **ACIE** in order to keep the available research in civil and industrial engineering as current and timely as possible.

## COVERAGE

- Operations Research
- Ergonomics
- Construction Engineering
- Urban Engineering
- Quality Engineering
- Optimization Techniques
- Production Planning and Control
- Transportation Engineering
- Earthquake engineering
- Hydraulic Engineering

IGI Global is currently accepting manuscripts for publication within this series. To submit a proposal for a volume in this series, please contact our Acquisition Editors at Acquisitions@igi-global.com or visit: http://www.igi-global.com/publish/.

The Advances in Civil and Industrial Engineering (ACIE) Book Series (ISSN 2326-6139) is published by IGI Global, 701 E. Chocolate Avenue, Hershey, PA 17033-1240, USA, www.igi-global.com. This series is composed of titles available for purchase individually; each title is edited to be contextually exclusive from any other title within the series. For pricing and ordering information please visit http://www.igi-global.com/book-series/advances-civil-industrial-engineering/73673. Postmaster: Send all address changes to above address. 

## Titles in this Series

*For a list of additional titles in this series, please visit:*
*www.igi-global.com/book-series/advances-civil-industrial-engineering/73673*

***Health and Well-Being Considerations in the Design of Indoor Environments***
Roberto A. González-Lezcano (Universidad CEU San Pablo, Spain)
Engineering Science Reference • © 2021 • 286pp • H/C (ISBN: 9781799872795) • US $215.00

***Visibilities and Invisibilities in Smart Cities Emerging Research and Opportunities***
H. Patricia McKenna (AmbientEase, Canada)
Engineering Science Reference • © 2021 • 289pp • H/C (ISBN: 9781799838500) • US $195.00

***Transforming Urban Nightlife and the Development of Smart Public Spaces***
Hisham Abusaada (Housing and Building National Research Center, Egypt) Abeer Elshater (Ain Shams University, Egypt) and Dennis Rodwell (Independent Researcher, UK)
Information Science Reference • © 2021 • 290pp • H/C (ISBN: 9781799870043) • US $195.00

***Handbook of Research on Driving Transformational Change in the Digital Built Environment***
Jason Underwood (University of Salford, UK) and Mark Shelbourn (Birmingham City University, UK)
Engineering Science Reference • © 2021 • 569pp • H/C (ISBN: 9781799866008) • US $295.00

***Structural Dynamics and Static Nonlinear Analysis From Theory to Application***
Mourad Belgasmia (Setif 1 University, Algeria)
Engineering Science Reference • © 2021 • 347pp • H/C (ISBN: 9781799843993) • US $195.00

***Handbook of Research on Developing Smart Cities Based on Digital Twins***
Matteo Del Giudice (Politecnico di Torino, Italy) and Anna Osello (Politecnico di Torino, Italy)
Engineering Science Reference • © 2021 • 674pp • H/C (ISBN: 9781799870913) • US $295.00

IGI Global
PUBLISHER of TIMELY KNOWLEDGE

701 East Chocolate Avenue, Hershey, PA 17033, USA
Tel: 717-533-8845 x100 • Fax: 717-533-8661
E-Mail: cust@igi-global.com • www.igi-global.com

# Table of Contents

# Detailed Table of Contents



The United Nations (UN) has proposed two actions against climate change between 2015 and 2021: "combat" in Goal 13 of the Sustainable Development Goals (SDGs) and "adaptation" in the 26th UN Climate Change Conference (the Conference of the Parties COP 26). This chapter aims to highlight pathways and actions for addressing and adapting to climate change at higher strategic levels and urban planning and design at the local level. In 32 authoritative texts, the snowball technique and content analysis were used to discover the interactions between people, nature, and climate change adaptation. The findings revealed that lower-level adaptation methods, such as urban design techniques, were ineffective in responding to people's actions in public areas. In terms of SDGs and COP 26, epistemological awareness of normative variables crucial to the relationship between people and nature in public spaces adds significantly to this endeavor.



Urbanisation affects the environment and ecosystems in many ways. For example, cities are continuous, resulting in their footprint being extended. Rapid urbanisation is driving cities to become significantly warmer than the surrounding countryside as a result of this rapid urbanisation and the unsustainable attitude of their citizens toward the city. The Sustainable Development Goals (SDGs) were adopted by countries worldwide a few years ago. This will enable challenges. In light of this, cities need to adopt sustainable urbanization, allowing citizens to participate in participatory, integrated, sustainable city planning and city management to enhance the quality of life. By implementing the SDGs, cities can help reverse the damage done to their ecosystems. As part of its 2052 National Vision, Egypt has also implemented the SDGs and achieved targeted improvements for its citizens. This chapter concludes with comprehensive initiatives in the context of implementing the SDGs.



This chapter discusses the historic areas that are living habitats that grew before the science of architecture and urban design happened in a formal way. In the case of the Kampung Kauman neighbourhood area, the area is a result of an agglomeration of residents with similar job types as Batik makers. However, the district was designed by an indigenous Javanese planner or Kalang who has a responsibility to design and build the area. The method used a combination of computational climate simulation to investigate outdoor spaces, surface radiation. Further, this chapter conducted an ethnographic study qualitatively through field surveys. The results captured people's activities at different times during the day. The results also indicated that most of the Javanese buildings can consider climate as responding to buildings and public spaces. The concluding remarks identified the relation of the urban morphology from indigenous Javanese culture with people activities through spatial investigation.



Solid waste management is considered one of the greatest challenges faced by historic centres due to the nature of their urban fabric and the changing frequency of users of the area, resulting in variability in the amount and composition of solid waste produced. Although historic centres are places where the Genius Loci of cities are appreciated, the implications of poor solid waste management led to the deterioration of these significant places. The application of holistic strategies contributes to keeping the urban identity alive while considering current social and economic aspects. This chapter focuses on exploring the potential of using the urban regeneration framework for solid waste management in historic centres. The discourse begins by defining the conceptual framework, followed by reflections on the historic centres of Naples and Cairo to explore how regenerative approaches can be used. Finally, the findings suggest a possible regenerative framework that could be used by different groups to solve problems with waste management in historic cities.



Global and regional warming phenomena drive the effect of climate change on cities. Research on urban heat adaptation strategies has been growing, with many studies concentrating on blue and green infrastructure (BGI) as an alternative to grey infrastructure and to minimise the adverse effects of urban heat islands. Building upon investigating the case of Germany, this chapter presents mitigation strategies for Alexandria, Egypt that can reduce the temperature at the urban design level and identify urban heat island effects through BGI and their performance, as well as their importance in achieving sustainable urban development and low-temperature cities. The lessons learned show that BGI is a new approach to city planning that can minimise the impact of human development on functioning natural systems, the quality of life, and the economic consequences. A set of BGI guidelines for Alexandria on how to keep cities cool and improve the city's environment were agreed upon.



This chapter applies a design-based approach to integrate the innovative greenery measures in a high-density unplanned area of Imbaba to mitigate the urban heat island (UHI) effect and provide high quality open spaces by creating a chain of cooling public spaces called "Takeeba." The literature review was applied to propagate the formulation of the research problems and identify the boundaries of knowledge. Land surface temperature (LST), the normalized difference vegetation index (NDVI), and the normalized difference built-up index (NDBI) were mapped using remote sensing to identify the hotspots and the existence of greenery and vacant lands for intervention. Prioritizing the hotspots shows vacant land and parking lots are the hottest spots due to direct exposure to solar radiation and heavily compacted areas in the centre of Imbaba. To tackle UHI in the whole Greater Cairo, the replicability of the idea was considered in a follow-up project.

This chapter examines a community developed in the early 20th century as a sustainable development paradigm. However, today saw the implementation of particular development plans in some cities in the global south that may not align with this community's long-term goals. Two Delphi rounds and an Envi-Met microclimate simulation helped researchers of the present work use Heliopolis as a case of investigation. After a Heliopolis' urban development plan competition, this approach's results show a considerable drop in the site's initial values and ecological effect. Surface and mean temperatures and carbon dioxide levels are predicted to decline because of this significant loss. Future studies can look at how local streets become urban boundaries and context segregation because of development plans that cannot regenerate urban areas.



Mobility and safety are the main characteristics of the city's urban strategies that regenerative cities adopt. This study is concerned with the mobility of pedestrians while crossing vehicle routes within urban developments. Several research groups recognise the negative-positive impact of ergonomic design in understanding human limitations and abilities; additionally, most researchers mentioned ergonomic design in the design of interior architecture elements, and a few have introduced it on an urban scale. So, research used ergonomic design to analyse pedestrian and vehicle movement. The research method adapted ergonomic design principles as a method to analyse human behaviour while crossing roads in an Egyptian context. After reviewing the relevant literature and Egyptian context, the authors discovered that studies show significant development occurred on Cairo's roads that threaten people's lives. The analysis tools used in research are questionnaires, surveys, and observations. The results concluded that both streets need more pedestrian-friendly designs and regulations.

Global research extensively discusses artificial intelligence (AI) and thermal comfort in urban studies. The current research design uses Goodvision Video Insights as an application of AI to analyse users' preferences for movement within three exits/entrances of the metro stations in Ramses Square, Cairo, Egypt. This study uses AI technology to track pedestrian movement based on heat exposure in metro station contexts. The primary findings identified the relationship between user experiences and urban composition. These results also revealed that the urban morphology of the metro station context impresses users' thermal comfort and movement preferences enough to track their needs through commutes. This study could not address all the aspects that influence users' experiences in the case study. Future studies provide light on the impact of varying temperature conditions on comfort levels and, as a result, human behaviour in station areas.



The analysis of pedestrian wind comfort was carried out using computational fluid dynamics (CFD) on a 3D model of the Istanbul Bilgi University Central Campus. The context of the analysis was conducted on the pedestrian activity maps within the campus between November 2021 and January 2022, including a time of strong south-westerly winds (Lodos) that hit Istanbul at the end of November/early December 2021. The activities of students and their movement patterns were observed through videos and by counting the pedestrians manually, creating a spatial map that represents pedestrian counts and their activities within the outdoor spaces. A 3D model of the campus was constructed, and CFD pedestrian wind comfort analysis was performed and compared to the pedestrian activity maps. Spatial relation analysis was performed to examine whether the pedestrian activities can be related to the Lawson wind comfort criteria map. The results show that there was generally some inconsistency between the empirically observed number of pedestrians and their activities and the wind comfort criteria map.

# Preface

Signs of interest in weather began in the second half of the nineteenth century, specifically between 1772 and 1864, in two volumes of the Englishman Luke Howard (2012) on meteorological observations in London. The impact of weather on agriculture and navigation was also studied. In the past decades, protecting the urban environment in the face of environmentalism and environmental rights has become crucial to saving the planet from the dangers of the rapid urban development of new cities and societies, mainly the digital and smart. The first beginnings of this argument have occurred since Rachel Carson (1962) published her book *Silent Spring* talking about environmental impacts, followed by two books of Ian McHarg (1996; 2006), *Design with Nature* and *Man and Environment*. Both works represent the early beginnings of a discussion on ecological balance while respecting the right of future generations to their share of natural resources. Their ideas led to the multiplicity of architectural and urban research to support environmental concerns and compatibility between people and the surrounding environment, which paved the way for introducing sustainable cities in urban planning and design (Yang & Li, 2016).

From the end of the last millennium to the present, this thought occupied a wide area of interest among professional theorists and practitioners in the field of city planning and design, and the focus of attention has become meeting the needs of the present without compromising the ability of future generations to meet their needs. Therefore, the idea of sustainability has expanded and has become a requirement for specialists, whether academics or practitioners, to reconcile the environmental, social, cultural, economic, and technical demands necessary for the survival of humanity and living creatures on planet Earth.

Sustainable urban development has taken the form of goals (Jones, Wynn, Hillier, & Comfort, 2017) and visions (United Nations, 2015). Accompanying each other is the correlation of the four natural components water, food, energy, and climate in integration with the equation of placemaking out of people and place (Abusaada & Elshater, 2021; Jones, Wynn, Hillier, & Comfort, 2017). Among these goals and visions, there were calls in the face of climate change and environmental degradation [Goal 13 of the Sustainable Development Goals (SDGs)] (Tang & Lee, 2016).

In the past decades, vegetation cover has declined because of climate change (Bachelet, Neilson, Lenihan, & Drapek, 2001). Soltani and Sharifi (2019) see that this shortage of vegetation cover "alongside urban structures" and "hardscape" is one of the reasons for increasing temperature in urban environments. This phenomenon is known as the "urban heat island (UHI) effect." Climate change is accompanied by a massive expansion of urban structures in densely populated urban areas. Thus, temperatures rose above normal levels in public spaces in all cities of the world, such as India (Kotharkar & Bagade, 2018) and Bangladesh (Dewana, et al., 2021) in Asia, Budapest, Hungary, in Europe (Diana, Pongráczab, Dezsőa, & JuditBartholyab, 2020; Buzási, Pálvölgyi, & Csete, 2021), and in the USA between urban and rural areas (Fu & Weng, 2018). Moreover, climate change has become a threat to sustainable urban development goals as it caused an increase in poverty and inequality, especially in African countries (Chirambo, 2017).

As a result, it has become a primary indicator of the measurement of the adaptive dimensions of social and cultural life and livelihoods (Adger, Barnett, Brown, Marshall, & O'Brien, 2013) and nature-based solutions in the face of socioeconomic forces (Lafortezza, Che, Bosch, & B.Randrup, 2018). Therefore, scientists have created many methods, tools, and techniques to map areas with elevated air temperature, which they called the UHI phenomenon. This book covers many techniques used to measure UHIs (Budhiraja, Pathak, & Acharya, 2019; Danahy, Mitchell, Wright, Hoinkes, & Feick, 2017; Ngarambe, Oha, Sua, Santamouris, & Yuna, 2021):

- Multi-scale 3D Geovisualization
- Multi-Sensor Data in Geographic Information Systems
- An analysis of variance and combined clustering techniques
- Space syntax approach
- Occurrence of heat waves
- Empirical and spatial ethnographical analysis
- *In situ* (field) measurements

This book reviews the fundamentals of this local climatic phenomenon as a gateway to solving the challenging problems of rapid urbanization in the face of climate change. This work attempts to use the dimensions and principles of urban planning and design, and landscape architecture in conjunction with the competence of environmental design to reduce the impact of this phenomenon. The book focuses on five SDGs to explain the problems that urban residents suffer because of high temperatures or the formation of heat islands. These selected SDGs are Goals 1, 3, 8, 11, and 13. Some of which can be limited to affecting the health status, productive capacity, social and economic well-being, and the feeling of distress and aggressive behavior.

## SYNOPSIS

Air temperature is one of the factors influenced by climate change and contemporary city morphology that lacks compact city features. This phenomenon is mainly related to the city characteristics of the urban spatial configuration at the level of the general plan of the city and the characteristics of materials and population density that increase the temperature of cities (Sangiorgio, Fiorito, & Santamouris, 2020). These cities had a spatial configuration that combined the elements of traditional cities in the urban fabric with low buildings, narrow roads, and public spaces. However, at the same time, this structural system achieved an acceptable thermal equilibrium throughout the year. On the other hand, in most of the cities of the world, contemporary cities have taken global paradigms, adopting open-fabric, multiple, and ultrahigh residential towers and superhuman scale-spaces at the level of squares and public parks. This kind of planning caused a radical thermal transformation in the network of movement and transportation and all public spaces and their external spaces and even on the level of residential streets and semipublic and semiprivate areas.

The term "urban heat islands" (UHIs) was revitalized to examine the contribution of cities to global warming. This new term is related to temperature differences between urban and surrounding rural areas (Kotharkar & Bagade, 2018). As rapid urbanization and new urban development projects have led to an increase in air temperatures, the creation of the spatial regions of the city is characterized by temperatures exceeding their normal levels. It also varied because of the difference in heat intensity between night and day (Dewana, et al., 2021). For example, the maximum effect of UHI in the central business district area in Adelaide in Australia occurs at midnight (Soltani & Sharifi, 2019).

Some causes of the UHI phenomenon can be limited to weather conditions and urban morphology (Ngarambe, Oha, Sua, Santamouris, & Yuna, 2021). However, the UHI phenomenon is also caused by concrete pavements near the rivers, embankment systems, and the enormous influx of traffic flow (Buzási, Pálvölgyi, & Csete, 2021), as well as abnormal surface covers in built-up areas and non-natural surface covers (Diana et al., 2020). UHIs also occurred because of the hyper-diversity of street worker activities and urban spaces of commercial and recreational shops. These UHIs have adverse effects on the vitality and joy in outdoor spaces and recreational areas, such as beaches, resorts, parks, sports stadiums, and public squares, affecting public health (Soltani & Sharifi, 2019), laziness, and social and economic well-being. These negative effects represented the lack of productive capacity resulting from excess heat stress, high poverty rates, and aggressive behavior characterized by the stampede of physical abuse, verbal abuse, and increased disturbance caused by vehicle alarms and pedestrian disturbance. Furthermore, UHIs can also be due to radiation from artificial lighting in public spaces (Liu, Xub, Weng, & Shu, 2021; Zhou, Chen, Wang, & Zhan, 2011).

The impacts of UHIs on people in urban environments are also described in this book. In urban environments with no UHIs, the well-being and vitality of people increased, as well as their physical activity and productive capacity; their morale is improved; and their behaviors are controlled. This book has two parts. The first part focuses on the concepts of UHIs caused by the increase of air temperature and solar radiation, air movement, and relative humidity. This part shows the digital methods and contemporary techniques used to measure UHI indicators and provide solutions to reduce their impacts on urban areas. The second part uses sustainable urban development goals to improve the diversity of the characteristics of urban environments. It focuses on changing urban form and everyday life experiences to neutralize the adverse effects of UHIs in modern cities, especially the smart and digital ones.

This book proceeds to show two main issues:

1. Regenerative cities and community well-being: In this issue, the syndromes of over-urbanization and the massive expansion of urban structures in densely populated urban areas caused by increased urban island areas are examined. To provide a set of planning and design solutions, the focus should be on achieving social and economic well-being, increasing productive capacity, reducing poverty, achieving equality among the members of society, supporting urban vitality, and improving the environment.
2. Challenges of micro and macro climate change in the changing world governed by technology and scientific progress: This issue attempts to present innovative ideas and concepts related to the UHI phenomena:
   - In situ (field) measurements
   - Surface UHI intensity
   - UHI scientific methods and techniques

In addition to models associated with "human thermal comfort" and making a place for people in the era of new, post, and next urbanism, these chapters depend on urban planning and design, landscape architecture, and environmental design based on all the issues mentioned above to create new sustainable, smart, and digital cities.

## AIMS AND OBJECTIVES

This book focuses on four SDGs: public health and well-being (Goal 3), decent work and economic growth (Goal 8), sustainable cities and societies (Goal 11), and climate action (Goal 13). These goals are associated with the increasing UHI phenomenon that accompanies rapid urbanization, which has changed the way of life of many countries worldwide. Thus, this book aims to reach sustainable cities and societies that do not suffer from poverty and disease due to climatic change and where decent work and social and economic well-being is achieved. The objectives of this book are as follows:

- To present the atlases of UHI-based digital techniques and methods of modeling.
- To use these atlases, mapping, and models in exploring the placemaking problems in the new cities, particularly in the sustainable, smart, and digital cities.
- To use all the above suggestions to present solutions to solve the five SDGs (Goals 3, 8, 11, and 13).

## TARGET AUDIENCE

The prime audience includes experts working in architecture, site planning and design, urban planning and design, landscape architecture, sustainable urban design, and environmental design. In addition, the book focuses on researchers, academics, practitioners, and urban governance, developers, and policymakers. Significantly, the target audience can get more insights into using new paradigms, methods, and research applications.

## TOPICS COVERED

This book has two groups of chapters with many recommended insights for climate changes, UHI effects and regenerative cities. In the first group of chapters (Chapter 1– Chapter 5) show the theoretical fundamentals of mapping and the impacts of UHIs. These chapters aim to interpret the phenomenon of UHIs in general and mainly in terms of the sustainable urban development goals for developing cities and societies. This part provides a reliable practical and scientific basis for causes, measurements, and the latest techniques in addition to devices and aims to interpret this phenomenon. It also discusses how academics, researchers, and practitioners, separately and collectively, use these technical issues to build the atlas to map UHIs for each city.

The second group of chapters (Chapter 6 – Chapter 10) explores how research uses the environment effects to explore SDGs in the Global North and South. These chapters focus on society and community problems, such as sociocultural, economic, and technological problems. In addition, this group articulate the placemaking problems, dealing with public health and well-being, decent work and economic growth, and sustainable cities and societies, aiming to make cities and human settlements inclusive, safe, resilient, and sustainable.

The many measuring, modeling, and mitigation strategies according to sustainable urban planning and design covered in this publication include the following:

- Daytime surface UHI intensity
- Decent work and economic growth
- Environmental design
- Exaggerating urbanism
- Geovisualization
- Landscape architecture
- Nature-based solutions
- Photogrammetry and remote sensing
- Remote sensing
- Satellite-based estimation
- Site planning and design
- Social and economic well-being
- Spatial analysis
- Street workers
- Sustainable design
- Well-designed public spaces

## THE BOOK CHAPTERS

Throughout the book, ten chapters cover each topic. Climate change adaptation is introduced in the first chapter by Elshater and Abusaada. The authors of this chapter begin by discussing two action plans proposed by the United Nations (UN) against climate change between 2015 and 2021: "combat" in goal 13 of the Sustainable Development Goals (SDGs) and "adaptation" at the 26th Conference of the Parties (COP 26). The objective of this chapter is to highlight different pathways and actions for addressing and adapting to climate change at a higher strategic level and urban planning and design at a local level. The snowball method and content analysis were used to uncover the connections between people, nature, and climate change adaptation in 32 authoritative texts. In terms of responding to people's actions in public areas, lower-level adaptation methods, such as urban design techniques, proved ineffective. As part of the SDGs and COP26, epistemological knowledge of normative variables is crucial to the relationship between people and nature in public spaces. There is no effective cooperation between the professional entities and the universities.

The second chapter, provided by Lina Fayed and Ahmed Fayed, examines the impact of urbanisation on the environment and ecosystems in various ways. According to them, cities, for example, are continuous, thereby extending their footprint. Cities are getting warmer than the surrounding countryside due to increasing urbanisation and a change in attitude towards the city. The authors of this chapter mentioned that a few years ago, countries worldwide adopted the Sustainable Development Goals (SDGs). As a result of their deductive argument, challenges are likely to arise. The results here conclude that cities should embrace sustainable urbanisation. To improve residents' quality of life, policies can enable participatory, integrated, and sustainable planning and management process. As part of its 2052 National Vision, Egypt has incorporated the SDGs and made targeted changes for its inhabitants. In addition to meeting the SDGs, the country made targeted improvements for its residents as part of its 2052 National Vision. The chapter concludes with comprehensive actions aimed at achieving the SDGs.

Chapter 3 by Melvina Pramadya Puspahati discusses Kampung Kauman as a historical neighbourhood that grew out of a concentration of people working in the same field as batik manufacturers. She mentioned that Kalang planners or indigenous Javanese planners were responsible for planning and building the district. To identify the relationship between indigenous Javanese culture's urban morphology and the activities of its inhabitants, a spatial investigation was conducted by the author of the chapter. In the study of outdoor environments, climate simulation and surface radiation measurements were combined. Here, aualitative field questionnaires that were used also by the author of that chapter in this chapter to conduct an anthropological study. Data was collected about the day-to-day activities of the people. The study also found that Javanese buildings can take into account the effects of climate on their surroundings.

In Chapter 4, one of the challenges involved in planning and constructing sustainable cities is environmental disposal of solid waste. Mariana Barrera and Yara Elmaghrabi describe the management of solid waste in historic centers due to the nature of the urban fabric and the changing frequency of users of the area. The authors examined the amount and nature of solid waste generated. Even so, the consequences of poor solid waste management lead to the decline of cities' Genius Loci, which are historic centers. By using holistic techniques, urban identity can be preserved while taking current social and economic factors into consideration. In that chapter, the feasibility of adopting the urban regeneration framework for solid waste management is explored. The discussion begins by defining the conceptual framework, then moves on to comments on the historic centres of Naples and Cairo to investigate how regenerative approaches may be applied. The study suggests an effective regenerative framework that could be used by different groups to alleviate the challenges associated with waste management in historic cities.

As part of mitigation of heat-related exposure, Riham Nady Faragallah recommended using green infrastructure in Chapter 5. According to the author, the effects of climate change on cities can be attributed to both global and regional warming. In order to reduce the negative effects of urban heat islands, blue and green infrastructure (BGI) has received considerable attention as an alternative to gray infrastructure. In Alexandria, Egypt, this chapter discusses urban design mitigation techniques that reduce temperatures. Additionally, urban heat island impacts are discussed through BGI and their performance, as well as their role in achieving sustainable urban development and cool cities. The BGI is a novel approach to planning that has the potential to minimize the impact of human development on natural systems, the quality of life, and the economy. The Alexandrians developed a set of BGI principles to keep cities cool and improve the environment.

We now proceed to Parisa Kloss and Heba Allah Essam E. Khalil in Chapter 6, in which they analyse the consequences of urban heat islands on informal settlements. This chapter discusses a design-based method for incorporating vegetation measures into an unplanned high-density region in Imbaba, Giza, Egypt, in order to mitigate the urban heat island effect and offer high-quality public spaces. As a result of the investigation, Takeeba, a chain of cooling public spaces, was chosen as the case study's building style. The literature review was used to define the knowledge boundaries. The authors of this chapter were able to map the land surface temperature (LST), the normalised difference vegetation index (NDVI), and the normalised difference built-up index (NDBI) using remote sensing in order to pinpoint hotspots and unoccupied lots. According to the analysis of the authors of this chapter, unoccupied land and parking lots have been identified as the areas of highest priority. This is due to direct solar radiation exposure and densely packed locations in Imbaba's centre. This chapter recommends future study that would examine whether the concept could be duplicated in the Greater Cairo Region.

After defining urban heat island, Chapter 7 looks at a sustainable community built in the early 20th century. Hamdy Hefnawy, Abeer Elshater, Mohamed El Fayoumi, and Natalie Marie Gulsrud presented this chapter. There is an examination of the impact of growth plans in some parts of the global south that may be at odds with the long-term goals of this population. A microclimate simulation using Envi-Met and two Delphi rounds were used to study Heliopolis. Following a competition for Heliopolis' urban development plan, this strategy resulted in a significant decrease in the site's initial values and ecological impact. As a result of this considerable loss, it is expected that surface and mean temperatures, as well as carbon dioxide levels, will decrease. Further research should explore the process by which neighborhood streets become urban boundaries. As a result of development plans that fail to revitalize urban areas, context segregation could also be examined in this study.

The following three chapters deal with regenerative cities and their environmental impact on pedestrians. In Chapter 8, Nardine El-Bardisy, Alaa N. Rashed, Eman Ahmed Elmeligy, and Hala Hussien Aboulnaga discuss Heliopolis's safety and mobility issues in Cairo. According to the authors of this chapter, mobility and safety are two primary characteristics of urban regenerative solutions. Their research focuses on pedestrian mobility when crossing vehicular routes in metropolitan areas. According to that chapter, several research groups acknowledge the beneficial and detrimental effects of ergonomic design on understanding human limitations and capabilities. Several researchers have mentioned ergonomic design in creating interior architecture elements, and some have implemented it large scale. Researchers then used ergonomic design to analyse pedestrian and vehicle movements. To analyse human behaviour on highways in an Egyptian context, ergonomic design principles were applied. Reviewing relevant literature and the Egyptian context, we observed that studies indicate significant development has occurred on Cairo's roads, posing a danger to human life. The methods of analysis used in research include questionnaires, surveys, and observations. The study found that both routes need more pedestrian-friendly design and regulation.

In Chapter 9, Shereen Wael, Abeer Elshater, and Samy Afifi examine artificial intelligence (AI) and thermal comfort based on a case study from Egypt. They used Goodvision Video Insights as an application of artificial intelligence to analyze users' movement preferences at three metro station exits and entrances in Cairo, Egypt. This work employs artificial intelligence to track pedestrian movement in metro station situations based on heat exposure. One of the key conclusions is that user experiences are related toition. The results also indicate that the urban morphology of the metro station surroundings significantly influences passengers' comfort and mobility preferences during transportation. In the case study, it was not possible to address all factors that affected users' experiences. Taking into account the limitations of this study, it is essential, according to the authors of this chapter, to shed light on the effects of different temperatures on human comfort and, consequently, on human behavior in station areas.

Finally, Abdalrahman T. Y. Alashi, Turgay Kerem Koramaz and Dilek Yildiz Ozkan discuss the computational fluid dynamics (CFD) used to analyse pedestrian wind comfort on a three-dimensional model of the Istanbul Bilgi University Central Campus. They conducted their analysis between November 2021 and January 2022, which included a period of strong south-westerly winds (Lodos) affecting Istanbul at the end of November and early December 2021. Students' activities and movement patterns were studied using video and manually counting pedestrians, which resulted in a map representing pedestrian counts and their activities in outdoor locations. CFD pedestrian wind comfort analyses were conducted along with pedestrian activity maps to create a three-dimensional model of the campus. Authors of that chapter conducted a spatial relationship study to see if pedestrian activities might be linked to the Lawson wind comfort criteria map. As a result of the analyses, there was some disagreement between the number of pedestrians and their activities observed empirically and the wind comfort map.

*Hisham Abusaada*
*Housing and Building National Research Centre, Egypt*

*Abeer Elshater*
*Ain Shams University, Egypt*

## REFERENCES

Abusaada, H., & Elshater, A. (2021). Effect of people on placemaking and affective atmospheres in city streets. *Ain Shams Engineering Journal*, *12*(3), 3389–3403. Advance online publication. doi:10.1016/j.asej.2021.04.019

Adger, W. N., Barnett, B., Brown, K., Marshall, N., & O'Brien, K. (2013). Cultural dimensions of climate change impacts and adaptation. *Nature Climate Change*, *3*(2), 112–117. doi:10.1038/nclimate1666

Bachelet, D., Neilson, R. P., Lenihan, J. M., & Drapek, R. J. (2001). Climate ehange effects on vegetation distribution and carbon budget in the United States. *Ecosystems (New York, N.Y.)*, *4*(3), 164–185. doi:10.100710021-001-0002-7

Budhiraja, B., Pathak, P. A., & Acharya, D. (2019). Studying surface and aanopy layer urban heat island at micro-scale using multi-sensor data in Geographic Information Systems. In *Geospatial intelligence: Concepts, methodologies, tools, and applications* (pp. 389–410). IGI Global. doi:10.4018/978-1-5225-8054-6.ch018

Buzási, A., Pálvölgyi, T., & Csete, M. S. (2021). Assessment of climate change performance of urban development projects – Case of Budapest, Hungary. *Cities (London, England)*, *114*, 1–11. doi:10.1016/j.cities.2021.103215

Carson, R. (1962). *Silent Spring*. Houghton Mifflin.

Chirambo, D. (2017). Moving past the rhetoric: Policy considerations that can make Sino-African relations to improve Africa's climate change resilience and the attainment of the sustainable development goals. *Advances in Climate Change Research*, *7*(4), 253–263. doi:10.1016/j.accre.2016.11.002

Danahy, J., Mitchell, J., Wright, R., Hoinkes, R., & Feick, R. (2017). Multi-scale 3D geovisualization of urban heat island data for planning dialogue in Toronto. In C. N. Silva (Ed.), *Natural resources management: Concepts, methodologies, tools, and applications* (pp. 570–591). IGI Global. doi:10.4018/978-1-5225-0803-8.ch026

Dewana, A., Kiseleva, G., Golam, D. B., Mahmud, I., Bhuian, M. H., & Hassan, K. (2021). Surface urban heat island intensity in five major cities of Bangladesh: Patterns, drivers and trends. *Sustainable Cities and Society*, *71*, 1–12. doi:10.1016/j.scs.2021.102926

Diana, C., Pongráczab, R., & Dezsőa, Z. (2020). Annual and monthly analysis of surface urban heat island intensity with respect to the local climate zones in Budapest. *Urban Climate*, *31*, 1–16. doi:10.1016/j.uclim.2019.100573

Fu, P., & Weng, Q. (2018). Variability in annual temperature cycle in the urban areas of the United States as revealed by MODIS imagery. *ISPRS Journal of Photogrammetry and Remote Sensing*, *146*, 65–73. doi:10.1016/j.isprsjprs.2018.09.003

Jones, P., Wynn, M., Hillier, D., & Comfort, D. (2017). The sustainable development goals and information and communication technologies. *Indonesian Journal of Sustainability Accounting and Management*, *1*(1), 1–15. doi:10.28992/ijsam.v1i1.22

Kotharkar, R., & Bagade, A. (2018). Evaluating urban heat island in the critical local climate zones of an Indian city. *Landscape and Urban Planning*, *169*, 92–104. doi:10.1016/j.landurbplan.2017.08.009

Lafortezza, R., Che, J., Bosch, C. K., & Randrup, B. (2018). Nature-based solutions for resilient landscapes and cities. *Environmental Research*, *165*, 431–441. doi:10.1016/j.envres.2017.11.038 PMID:29217218

Liu, Y., Xub, Y., Weng, F., & Shu, W. (2021). Impacts of urban spatial layout and scale on local climate: A case study in Beijing. *Sustainable Cities and Society*, *68*, 102767. Advance online publication. doi:10.1016/j.scs.2021.102767

McHarg, I. L. (1996). *Design with nature*. Doubleday/Natural History Pressvith.

McHarg, I. L. (2006). *Man and environment* (F. Steiner, Ed.). Island Press.

Ngarambe, J., Oha, J. W., Sua, M. A., Santamouris, M., & Yuna, G. Y. (2021). Influences of wind speed, sky conditions, land use and land cover characteristics on the magnitude of the urban heat island in Seoul: An exploratory analysis. *Sustainable Cities and Society*, *71*, 102953. Advance online publication. doi:10.1016/j.scs.2021.102953

Sangiorgio, V., Fiorito, F., & Santamouris, M. (2020). Development of a holistic urban heat island evaluation methodology. *Scientific Reports*, *10*(17913), 1–13. doi:10.103841598-020-75018-4 PMID:33087829

Soltani, A., & Sharifi, E. (2019). Understanding and analysing the urban heat island (UHI) effect in micro-scale. *International Journal of Social Ecology and Sustainable Development*, *10*(2), 14–28. doi:10.4018/IJSESD.2019040102

Tang, H.-T., & Lee, Y.-M. (2016). The making of sustainable urban development: A synthesis framework. *Sustainability*, *8*(5), 1–28. doi:10.3390u8050492

United Nations. (2015). *Transforming our world· the 2030 agenda for sustainable development.* New York: UN.

Yang, B., & Li, S. (2016). Design with nature: Ian McHarg's ecological wisdom as actionable. *Landscape and Urban Planning*, *155*, 21–32. doi:10.1016/j.landurbplan.2016.04.010

Zhou, J., Chen, Y., Wang, J., & Zhan, W. (2011). Maximum nighttime urban heat island (UHI) intensity simulation by integrating remotely sensed data and meteorological observations. *IEEE, 4*(1), 138–146. doi:10.1109/JSTARS.2010.2070871

# Acknowledgment

The editors wish to express their gratitude to everyone who contributed to the production of this publication. We would like to express our sincere appreciation to the authors of the chapters in the current book for their active participation in the project. Thanks are also due to the Editorial Advisory Board for their constructive recommendations during the review process. Their dedication is greatly appreciated.

Josh Christ and Rowan Rumbaugh, Assistant Development Editors, Book Division, IGI Global, were invaluable in the development of this book. We are grateful to the Publisher IGI for providing us with a platform to inform our professional audience of the innovative views and thought-provoking arguments offered by the contributors to our publication.

In addition, the editors wish to acknowledge the Science, Technology & Innovation Funding Authority (STDF) for funding this book project [Grant number: STDF-BARG 37234].

# Chapter 1
# Climate Change Adaptation:
## Strategic Planning and Urban Design Practice

**Abeer Elshater**
https://orcid.org/0000-0002-5061-6861
*Ain Shams University, Egypt*

**Hisham Abusaada**
https://orcid.org/0000-0001-6530-7714
*Housing and Building National Research Center, Egypt*

## ABSTRACT

*The United Nations (UN) has proposed two actions against climate change between 2015 and 2021: "combat" in Goal 13 of the Sustainable Development Goals (SDGs) and "adaptation" in the 26th UN Climate Change Conference (the Conference of the Parties COP 26). This chapter aims to highlight pathways and actions for addressing and adapting to climate change at higher strategic levels and urban planning and design at the local level. In 32 authoritative texts, the snowball technique and content analysis were used to discover the interactions between people, nature, and climate change adaptation. The findings revealed that lower-level adaptation methods, such as urban design techniques, were ineffective in responding to people's actions in public areas. In terms of SDGs and COP 26, epistemological awareness of normative variables crucial to the relationship between people and nature in public spaces adds significantly to this endeavor.*

DOI: 10.4018/978-1-6684-2462-9.ch001

## INTRODUCTION

In September (2015), the United Nations (UN) adopted climate action[1] as a Sustainable Development Goal (SDG) 13 tactic to combat climate change and its impact[2]. Six years later, based on the findings of the United Nations Environment Programme (UNEP) (2021) on the gap in climate adaptation[3] (UNEP, UNEP DTU Partnership, World Adaptation Science Programme (WASP), 2021), on November 4, 2021, in Glasgow, England, the 26th UN Climate Change Conference (Conference of the Parties, COP 26) focused on climate change adaptation actions to build a sustainable world and better life for poor urban communities[4] (Moosmann et al., 2021). Adjusting physical, biophysical, biodiversity, social, political, economic, social, and ecological systems in response to climatic effects and stimuli was the focus of this adaptation (Kabisch, Korn, Stadler, and Bonn, 2017; Letcher, 2021; Lovejoy and Hannah, 2019).

In the 6 years between 2015 and 2021, climate change response strategies followed two paths that focused on taking measures related to climate change, one focusing on combat, (Letcher, 2021; Lovejoy and Hannah, 2019) and another on adaptation (Bovill, 2015; James, 2015), with different levels of control and adaptive responsibilities (or coexistence). The first path focuses on higher levels of development, such as nations and governments (a top-bottom approach) and specializes down to the level of individuals (a bottom-up approach). In contrast, actions in the second path concern individuals and progress all the way to governments and nations. This chapter deals with local-scale ideas for control, focusing on coping strategies, mainly in public places in poor communities in the Global South. According to Tabari and Willems (2018), there is considerable variation in seasons with respect to the traditional four distinct seasons. Das (2019) observed that these events of COP significantly impact people's comfort and lifestyles in the Global North and Global South.

*Climate change* is a dynamic driver of changes that leads to changes in urban policies. The overwhelming majority of scientists is convinced, based on scientific evidence, that humans are the sole drivers of modern climate change (Fares, 2021; Letcher, 2021). With regard to climate change and environmental crises, changing air temperatures, the intensity of solar radiation, humidity, torrential rains, floods, and storms are all caused by humankind's exploitation of Earth, which has led to noticeable consequences in terms of global changes (Letcher, 2021) in precipitation, frequency and magnitude of storms and hurricanes, floods, and droughts (Fares, 2021). In addition, the World Meteorological Organization (WMO) (2021) stated that because heat is becoming more intense in many regions, scorching summers are becoming more common.

Thermal comfort, which is affected by the regulation of urban heat islands (UHIs), is one of the primary subjects to be explored urban planning and design disciplines.

On a micro-scale, Soltani and Sharifi (2019) described a lack of planting cover and an increase in hardscape, which increases the UHI impact (i.e., in the form of an artificial temperature). For instance, the role of urban fabric on the surface of UHIs in Istanbul was depicted by Okumus and Terzib (2021). Elshater et al. (2022) also concluded that people in a district in Cairo, the capital of Egypt (particularly people who spend their time working and commuting outdoors), suffer from contextual effects of heat exposure.

Policy makers in the Global North can mitigate many of these influences of climate change policies, while countries of the Global South, particularly the poorer ones, have become overly affected by these policies. Consequently, experts worldwide have called for adaptation strategies to changing climatic conditions so that people can live better lives. In addition, there has been a growing call to study the economic realities in cities of the Global South, particularly the weak ones with limited financial capacity, and to address the dearth of low-cost technical solutions that make adaptation methods widely available (Bhan, Srinivas, and Watson, 2018). The gap between the North and South needs to be investigated. In this regard, control measures at the city level appear particularly important, they have not yet reached the small-scale public spaces in built environments. This level focuses on the effects on urban form and people's relationships with places. However, local level solutions have not been determined to cope with the results of these active interrelationships.

This chapter aims to develop a conceptual framework to highlight climate change mitigation and adaptation methods by determining the consequences of development plans and urban design practices, focusing on individual efforts at the local level. This chapter makes a significant contribution by summarizing the lessons learned at three levels: strategic planning, professional practice, and individual activities. This study also focuses on personal and local government participation to evaluate how many adaptation solutions are missing in urban planning and urban design for public places.

## BACKGROUND

The efforts of English-born painter, writer, philosopher, and art critic John Ruskin (1819–1901) led to the evolution of human-nature interactions in the nineteenth century. Ruskin (1848) was a firm believer in 'true to nature,' portraying natural beauty through mountains, oceans, sky, and hills, which he thought to be among nature's most attractive components. He formed a guild to protect the majority's rights and tried to make England a better place to live by viewing life as true wealth. Ruskin's book *Nature Studies*, revised and reprinted in 2016, contained excerpts from his environmental writings, in which he integrated nature and architecture within the context of an emphasis on social justice (Ruskin, 2016).

Much later, Frank Lloyd Wright (2012) introduced the most important interaction between people and nature in architecture in the 21st century. Since 1908, he promoted the concept of organic architecture, thinking that place character has an immediate impact on people's lives. He believed in organic architecture wedding the Earth, that the architect should be as creative as nature, and that buildings should be closely connected with nature and harmonize with the conditions of the surrounding environment. The influence of this thought was evident in designs inspired by the forms and principles of nature, such as The Millard House (1923), The Hanna House (1937), Taliesin (1911–1959), and Fallingwater (1964) (Payne & Gordon, 2019).

*Silent Spring*, by Carson (1962), is regarded as the birth of a modern environmental movement. It created the concept of "friends of the Earth," cautioned against environmental invasion, and sparked the creation of ecological policies and ideals. Ian McHarg (1969) published *Design with Nature* to support current environmental movements. McHarg is regarded as one of the modern era's fathers of sustainable development and believes that addressing natural processes is the best way to cope with human growth. He discussed an approach that advocates protecting and interacting with the current environment in all its forms without exposing it to any alteration, as was the case with the post-World War II reconstruction movement and the modifications it imposed that devastated its natural environment.

In the twenty-first century, literature examined the interaction between people and nature through the lens of global environmental change. Lomborg's (2001) literature, *The Skeptical Environmentalist*, came to be a second milestone in the trend towards ecological changes on Earth that affect the welfare of human beings, followed by a group of literature on climate change and its impact on the planet. The World Future Council developed the term "regenerative city" between 2008 and 2011 to describe the human-nature relationship in terms of methods for a more environmentally friendly, restorative relationship between cities and the ecosystems from which they draw resources for survival (Girardet, 2014). Developing comprehensive political, financial, and technological initiatives is a priority in regenerative cities. Its goal is to improve the compatibility between cities and the ecosystems from which they draw resources for survival (Girardet, 2014). Following that, a plethora of literature products related to the urban planning and design fields of specialization emerged, including *Urban sustainability in theory and practice: Circles of sustainability* (James, 2015), *Sustainability in architecture and urban design* (Bovill, 2015), *Planning sustainable cities and regions: Towards more equitable development* (Chapple, 2015), *City futures in the age of a changing climate* (Fry, 2015).

## METHODS

What are the viable solutions for climate adaptation at the urban planning and design level? How are these adaptive solutions applied to poor communities in cities of the Global South? To answer these questions, this chapter describes the sequential steps of the conceptual content analysis approach to identify the evidence of knowledge that answers these two research questions. Data mining was performed on two search streams: Google Scholar and Scimago Journal & Country Rank (SJR) with reference to the period from 2015 and 2021. In the former stream, the snowballing technique was employed to identify relevant references that focused on aspects of people-nature relationships. In the latter, data published in journal articles were investigated (contributions in conference proceedings, chapters, and books were excluded). Two main keywords were used for the search: "climate change" AND "sustainability". Secondary keywords also were used, namely: "sustainable development" OR "sustainable urban design" OR "sustainable urban planning" AND "regenerative" OR "adaptation" OR "urban heat islands". Data mining revealed 14 books published between 2015 and 2021 (Figure 1 and Table 1 in Appendix 1), as well as 18 articles from 16 journals in the SJR (Figure 2 and Table 2 in Appendix 1).

Data processing relied on the conceptual content analysis approach for the full text of the yielded results. The analysis looked for key terms and themes within qualitative data. In manually coding the samples, the existence and frequency of concepts relevant to climate change and UHI phenomena were investigated in the text. With manual coding, errors such as typos and misspellings could be easily recognized.

## ADAPTATION PATHWAYS

In the 6 years between 2015 and 2021, the use of keywords such as adaptation, climate adaptability, climate change, climate change adaptation, and UHI evolved through two phases, concentrating on distinct climate change-related measures, namely combat (taking urgent actions) (MCampbell, et al., 2018), and adaptation (nature-based solutions) (Frantzeskaki et al., 2019; Wamsler et al., 2020). Both phases have well-established principles in facing environmental problems, which are critical with two pathways of control and adaptive responsibilities (or coexistence).

1. The first path is geared towards higher administration levels such as nations and governments. It aims to combat climate change through renewable energy (Spyridi, Vlachokostas, Michailidou, Sioutas, and Moussiopoulos, 2015), anticipatory adaptation (Shearer, Coiacetto, Dodson, and Taygfeld, 2016),

*Figure 1. Keywords in 14 books between 2015 and 2021*

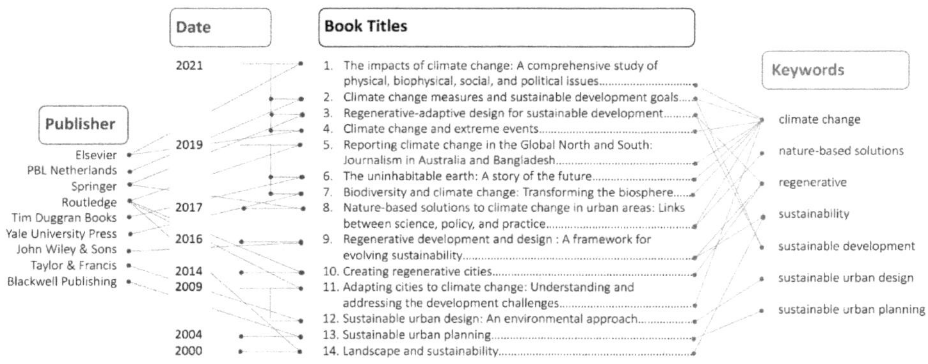


*Figure 2. Keywords in 16 journals and 18 articles between 2015 and 2021*

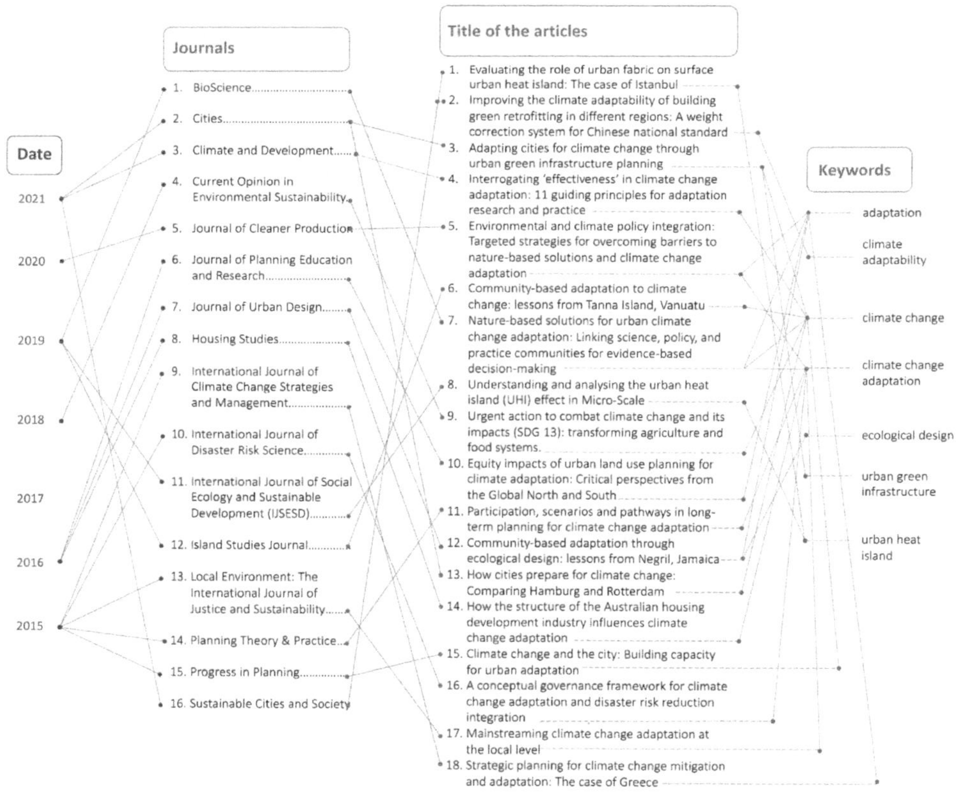

long-term planning (Campos, et al., 2016), institutional eco-innovation (Huang-Lachmann and Lovett, 2016), integrated nature-based solutions (Kabisch et al., 2017; Frantzeskaki et al., 2019), and the transformation of biodiversity and the biosphere (Lovejoy and Hannah, 2019).

2. The second path focuses on lower levels of decision-making, beginning with individuals and advancing to higher levels up to governments and nations (Rauken, Mydske and Winsvold, 2015). This path considers community-based adaptation (Dhar and Khirfan, 2016; Clarke, McNamara, Clissold, and Nunn, 2019; Singh et al., 2021) and participatory action research (supporting decision-making and collective action) (Campos et al., 2016).

Urban development considers the human-nature link and encompasses characteristics that emerge from six concepts: adaptation, compatibility, livability, regenerative capacity, resilience, and sustainability. Figure 3 shows our conceptual framework, which includes two-path suggestions for a climate-based solution. The first step explains the UN's role in climate change action, the second step is the literature review, and the third step includes our two paths for solutions in the two levels of adaptation: (I) development and strategic planning, and (II) professional practice and people's actions. The following review aims to track people-nature relationships regarding regenerative and adaptive capacities.

## Regenerative Capacity of Cities

This literature review incorporates principles such as designs with nature, respect for society's culture, and the use of hierarchy when making decisions regarding natural preservation, urban conservation, and regeneration. The products of the regenerative development approach are an ecosystem and a sustainable economy (Regenesis Group, 2016). Roös (2021) evaluated a regenerative-adaptive pattern-language theory that integrates regeneration and adaptability in design to achieve sustainable development. She based prospective solutions for significant environmental and sustainability concerns on the interconnectivity of "human and natural systems." Addressing the needs of behavioral patterns that recognize the duality of "man and nature." By supporting the living process, thinking in systems, working collaboratively, applying integrity in leadership and research, and embracing environmental care, the term "regenerative cities" makes regeneration systems fairer for future generations.

## Adaptive Capacity

Adaptation allows humankind to deal with environmental and climate change, particularly in local plans. This occurs in advance of the impacts and responses to

*Figure 3. Suggested conceptual framework based on two paths for solution in three steps*

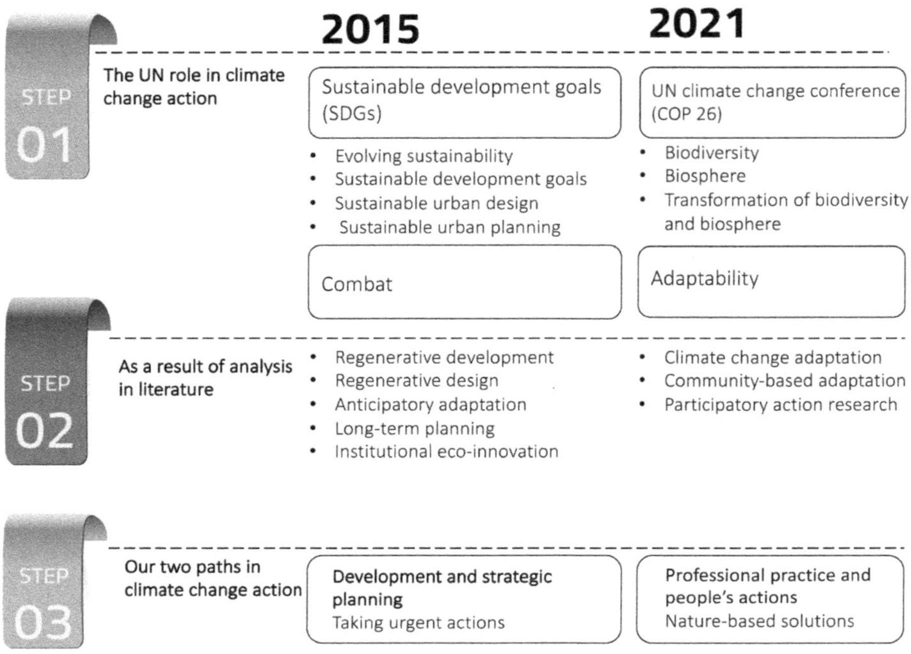


them. At the local level, adaptation focuses on stimulating actions among diverse stakeholder groups (Campos et al., 2016). In urban planning, urban design, and landscape architecture, adaptability is a need that is compatible with man-site interactions. This concept focuses on the interactions between humans and natural systems in their environmental surroundings (Roös, 2021).

In the context of climate change, Ramyar et al. (2021) explored adaptation as a guiding method in planning to achieve long-term sustainability and resilience as well as increase system resilience. They called for exploring ways to improve air quality and thermal comfort, manage food security and disasters, public health, biodiversity, and infrastructure longevity, among other facets of climate adaptation.

*Adaptive capacity* is a condition that enables people to foresee and respond to change and to recover from and minimize the repercussions of change (Ramyar, Ackerman, and Johnston, 2021). This criterion assesses the extent to which the human goal of developing an urban pattern in which the site's environment is intimately tied to the human person has been realized. It should serve as a repository of logic, foundations, and production laws. Urban planners and designers should link a location to the surrounding environment. In addition, each location should crystallize an appropriate cognitive formulation of the environment in which humans live and work. The issue of this criterion can be simplified to recognizing the limits of human and societal capacities to meet his requests within the natural and artificial variables that impact his decision to live where he or she wishes.

## STRATEGIC PLANNING AND URBAN DESIGN PRACTICE

As cities grew, their parts grew, and their activities proliferated over time. A collection of everyday tasks fulfilled various goals and requirements to make the city function efficiently, with various visuals and entrances. For example, 18 factors were considered in the two paths when combating and adapting cities to climate change from 2015 to 2021.

The first path is related to development and strategic planning. This path includes:

1. At the policy and planning levels, climate change adaptation, catastrophe risk reduction, and sustainable development are incorporated into government activities and decision-making (Forino, Meding, and Brewer, 2015).
2. By reducing $CO_2$ and other greenhouse gas emissions, increasing renewable energy sources and better forest management (i.e., tree planting and water) help to limit the rate of global warming (tree planting and water) (Spyridi et al., 2015).
3. Encourage institutional eco-innovations by providing a platform and incentives. (Huang-Lachmann and Lovett, 2016).
4. Improving natural systems provides climate mitigation, resilience, and adaptation planning options (Frantzeskaki et al., 2019).
5. Integrating environmental and climate policies: removing roadblocks into natural-based solutions and climate change adaptation (Wamsler et al., 2020).
6. Transdisciplinary approaches to unite multiple actors' efforts and capacities must integrate nature-based adaptation planning into municipalities' daily planning activities and the accompanying governance (Wamsler et al., 2020).
7. Reduced vulnerability and increased adaptive capacity are two ways to boost resilience (Dagnachew et al., 2021).
8. To reduce heat stress, improve the well-being of informal workers by modifying work hours and implementing behavioral changes (Singh et al., 2021).

The second path is related to the professional practice and people's actions. Here are some suggested actions for this path:

1. Exploring new approaches to understand and react effectively to urban adaptation challenges and opportunities through adaptation among city governors and planners (Carter et al., 2015).
2. Incorporating climate change adaptation into local development plans, local governments should incorporate adaptability into decision-making, policy creation, and service planning (Rauken et al., 2015).
3. Spatial planning can help in proactive cross-cutting urban adaptation by 'repositioning' it to emphasize ties to the core urban agenda, with a focus on collaborative approaches as a critical issue in meeting the challenge of urban adaptation (Carter et al., 2015).
4. Promote context-specific design through landscape urbanism, while also advancing proactive adaptation through ecological design that incorporates inputs from various professionals, including planners, environmentalists, climate change experts, and locals' nuanced knowledge and experience (Dhar and Khirfan, 2016).
5. Avoid making adaptation a privileged environmental good with exclusionary and maladaptive externalities by considering the distribution of adaptation benefits, costs, and responsibilities across society and treating all community groups equally, regardless of financial level. In addition, limiting the degree to which a system (vulnerability) is vulnerable to the negative consequences of climate change and seizing potential opportunities are both important (Campos et al., 2016).
6. Analyzing community contexts to recognize elements that have the potential to limit success and sustainability (Clarke et al., 2019).
7. Enhancing effective means to reduce temperature by articulating the UHI effects in simulating city energy use on a micro scale (Soltani and Sharifi, 2019).
8. Consider the contribution of urban fabric to microclimate issues and implement UHI abatement measures for projects to create climate-sensitive urban environments (Okumus and Terzib, 2021).
9. Enhancing urban green infrastructure through bottom-up actions, initiatives and participative integrated planning using nature-based practices (Ramyar et al., 2021; Vaňo et al., 2021).
10. Materials, building envelopes, and site design are all top priorities.

## DEDUCTIVE ARGUMENTS

The findings in this chapter (apparent in 18 factors) support our notion that the role of urban planners and designers in climate change adaptation at the level of public spaces is still not considered. Dhar and Khirfan (2016) observed that physical planning and urban design interventions are largely overlooked in this regard. Similarly, Anguelovski et al., (2016) believe that this lack of attention is particularly evident in poor communities in the Global South. Campos et al., (2016) demonstrated that disadvantaged and vulnerable groups in poor communities receive less attention.

The problem appears to be the insignificant temperature variations in public spaces between winter and summer, with few solutions available to handle natural seasonality. This issue is most visible in urban planning and design, ignoring climatic fluctuations throughout the year, day and night, public space patterns, and natural and landscape architectural features (Dhar and Khirfan, 2016; Roös, 2021). Furthermore, local inhabitants had little practical engagement in dealing with these changes. As in previous studies, another point worth mentioning is that, despite all the rhetoric regarding sustainability, livability and now, management and adaptation, there are no objective criteria to handle natural seasonal fluctuations and variations (Das, 2019; Tabari and Willems, 2018). This could be due to the fact that these are instantaneous changes that occur and disappear without monitoring and analysis by local institutional bodies in some cities.

Alternatively, climate change adaptation may have a noticeable influence on the local communities; however, the percentage of damage does not imply that the institutional mongering process is considered equally between countries. Rainwater floods the streets for several days in certain poor countries, yet little action is taken to drain the water because the expense of infrastructure development is prohibitive. Due to climate change, snowfall may occur in for 1 or 2 days. As such, no action that could have significant economic effects is anticipated within this short period. Finally, in many cities worldwide, mainly in the Global South, people suffer from environmental incompatibility[5].

Conflicts between climate adaptation and public space planning and design should be viewed as a better way to preserve city singularity (Abusaada, 2017; Elshater, Abusaada, and Afifi, 2019). This way is intended to expand options for innovative and efficient planning to utilize natural and human-made resources in the present and future. Such efficient planning focuses on an in-depth and mindful understanding of ecosystems. Furthermore, it identifies the most critical aspects of the natural environment and addresses them as a fundamental norm and starting point for every region's rehabilitation. It also helps reduce spatial isolation and strengthen the qualitative or quantitative link between urban space and the surrounding structures. In addition to facilitating the possibility of benefiting from contemporary developments, technologies, and services without prejudice to the state of the local natural and heritage environment and achieving socio-economic balance resulting from excessive economic growth due to increasing local capabilities and energy.

Climate change adaption strategies should focus on the everyday lives in cities and urban form. Such focus will also emphasize the use of traditional vocabulary in a contemporary form that meets the needs of today's population, bolstering aspects of a sense of place and familiarity and an interest in improving service efficiency and developing cognitive-behavioral awareness in individuals and groups. Strategies can consider public interest in environmental relations, the importance of maintaining cultural heritage, and working on environmental protection legislation.

Research centers should be encouraged to develop appropriate methods for preserving the natural environment and benefiting from its available resources without harm in order to study societal transformations in the resident group's thoughts and orientations and introduce customs, traditions, and folklore that enrich and raise the status of the individual and society. These concerns guide the values that regulate people's behavior, and the media, publications, and documents back them up to demonstrate their significance.

## URBAN ADAPTIVE SOLUTIONS

*Adaptability* is a criterion that can help shape and control the basis for choosing, planning, and designing public places in a city. Planners and designers view the concepts of people and nature's relationships in two dimensions.

***The first dimension*** tracks the function of a site when choosing a space to live in a new community. Adaptability depends on meeting place requirements and activities that are fulfilled in line with the city's expected function. For example, the place could be a city or an urban settlement that arises in a valley, desert, or even near existing cities. The city may have an industrial, tourist, or agricultural economic base, or it may be a target for housing, and it may group all or some of these functions.

***The second dimension*** is the planning approach to the site: the contribution to influencing the formulation of the foundations for choosing the site comprises two essential elements. Natural forces are the first elements, such as soil patterns, topography, wildlife resources (plants and animal families), surface and groundwater, and climatic influences. The forces formed by the civilized framework surrounding the place are the second element, including societal forces, cultural, economic, political, legislative, organizational, technological, and the exchange of information.

Hence, to select a site for new development projects, the most critical requirements of urbanization lie in human needs that are achieved by considering the relationship between people and their natural environments. In contrast, there must be consistency between human additions and natural conditions that determine the site's characteristics, leading to the site's recognition factor or surrounding context, which is one of the most important beginnings in dealing with the site from an adaptability standpoint. Adaptability focuses on searching for the most appropriate public places for these limited activities to meet the need for compatibility in its best aspects and better living-related thermal comfort.

Furthermore, numerous natural and manufactured treatments that can achieve acceptable urban adaptive solutions have been committed to developing adaptive capabilities. Soltani and Sharifi (2019) explained how a lack of vegetation cover increases the urban near-surface temperature at the microscale of the Adelaide metropolitan area in South Australia. In Istanbul, Okumus and Terzib (2021) concluded the same result regarding how compact urban fabric reduces land surface temperature and mitigates the effects of UHIs.

These paradigms should build public spaces based on new insights to achieve better placemaking and people satisfaction (Abusaada and Elshater, 2021b; Abusaada and Elshater, 2021c), focusing on planning and designing public places related to the relationship between urban form and everyday lifestyles on the one hand, and socio-economic, typo-morphological, environmental, spatio-temporal, and technological aspects on the other hand. In landscape architecture, hardscapes (stone, bricks, concrete, or metal) are used for landforms, pavements, and street furniture, while soft-scapes (planting or water) support thermal comfort and beauty and provide better visual impressions (Ramyar et al., 2021).

Accordingly, urban planners and designers must explore urban paradigms that adapt to the environmental dimensions of climate change adaptation, both at the level of vegetation cover and the pattern of urban fabric, for instance:

1. building characteristics: shape, height, size, material, profile, and configuration;
2. circulation networks: patterns, types, hierarchy, and pavements;
3. urban density: low, medium, or high;
4. urban green infrastructure planning.

Adaptation to climate change in built environments is more complex than in new-development areas in some global cities and impoverished areas (Anguelovski et al., 2016). People spend considerable time in public spaces in environments designed for work, mobility, and leisure. They have difficulty adapting to outdoor spaces that meet their needs and capabilities. Therefore, improving the adaptation pathways undertaken by these people requires a great deal of follow-up and field guidance.

1. Ceasing the construction of buildings, infrastructure, uses, activities, and materials that raise UHIs and affect thermal comfort.
2. Activating community participation between local authorities and representatives of individuals in each residential area.
3. Increase aid from NGOs and financial institutions to provide improvement supplies.

4. Benefiting from university and institute students' efforts to supervise the improvement of popular initiatives and develop feasible and straightforward designs (disassembly and installation, as needed) in their design studios.
5. Encouraging skilled workers in each area to participate in the improvement effort, either voluntarily or in return for an appropriate amount of money, as desired.

## CONCLUSION

The interrelationships between Mother Nature's impacts and the actions taken to combat climate change on our planet are active and could applicable. The outcomes of this cannot be ignored or read separately regardless of the distance between these impacts and actions. These relationships should be developed to display a roadmap that governs these outcomes in open and transparent cooperation driven by scientific research parties and based on appropriate solutions to deal with changing and complex situations. Furthermore, regardless of the level of SDG 13, the Paris Agreement, or even the UN COP 26 Conference, this cooperation did not generate any recommendations at the lower levels of urban planning and urban design. No feasible solutions for urban spaces in poor areas have emerged at the global level.

Following climate combat, adaptation is the second most significant factor to consider, and this chapter describes two paths to deal with continued climate change. Risk identification and assessment, environmental vulnerability discovery, natural system improvement, and institutional eco-innovation are the most important components of climate adaptation strategies, based on development planning as the first path. The central issue in the second path is people's participation in initiative-taking adaptation by presenting their contributions. Simultaneously, critical changes should be considered to address practical public space planning and design concerns. To adapt to climate change, the role of urban structure and resident behavior should be emphasized. A local community, particularly residents of a specific location alongside local authorities and non-governmental organizations, should be involved in developing an adaptation strategy plan.

This chapter explored the concept of climate change adaptation and identify what urban planners and designers do in this field, for instance with respect to public places in poor cities. The main contribution of this study lies in the recognition that adaptive elements in local areas require further research and testing. We suggest delving deeper into future adaptations through more in-depth critical analyses of pioneers' ideas regarding the compatibility between people and nature. In addition, planning and design for climate adaptation could be done through a learning and training process for residents. Social media may be an efficient means to educate people regarding adapting to climate change in public places. Developing the skills of residents of all ages and genders represents a core element for future research on improving the adaptability of public places. In addition, educating people combating climate change might be vital in future research.

## Conflict of Interest

The authors of this chapter declare there is no conflict of interest.

## Funding Agency

This research was supported by Science and Technology Development Fund (STDF), Egypt [Grant number: 37234].

## ACKNOWLEDGEMENT

The authors would like to thank the reviewers who provided time and effort in reviewing the previous version of this manuscript. We would like to thank Editage (www.editage.com) for English language editing. The authors also acknowledge the Science and Technology Development Fund (STDF), Egypt for funding this study of urban morphology and urban form [Grant: STDF Basic and Applied Research Grants (STDF-BARG: 37234].

## REFERENCES

Abusaada, H. (2017). The untold story of the urban design practice of new cities in Egypt. In H. Elsharkawy, S. Zahiri, & J. Clough (Eds.), *In the proceeding of the international conference for sustainable design of the built environment* (pp. 787–798). SDBE.

Abusaada, H., & Elshater, A. (2021a). Building sustainable habitats for free-roaming cats in public spaces: A systematic literature review. *Journal of Applied Animal Welfare Science*, 1–14. doi:10.1080/10888705.2021.2014840 PMID:34957863

Abusaada, H., & Elshater, A. (2021b). Effect of people on placemaking and affective atmospheres in city streets. *Ain Shams Engineering Journal*, *12*(3), 3389–3403. doi:10.1016/j.asej.2021.04.019

Abusaada, H., & Elshater, A. (2021c). Improving visitor satisfaction in Egypt's Heliopolis historical district. *Journal of Engineering and Applied Sciences (Asian Research Publishing Network)*, *68*(19), 1–22. doi:10.118644147-021-00022-y

Anguelovski, I., Shi, L., Chu, E., Gallagher, D., Goh, K., Lamb, Z., Reeve, K., & Teicher, H. (2016). Equity impacts of urban land use planning for climate adaptation: Critical perspectives from the Global North and South. *Journal of Planning Education and Research*, *36*(3), 333–348. doi:10.1177/0739456X16645166

Bhan, G., Srinivas, S., & Watson, V. (Eds.). (2018). *The Routledge companion to planning in the Global South*. Routledge.

Bovill, C. (2015). *Sustainability in architecture and urban design*. Routledge.

Campbell, B., Hansen, J., Rioux, J., M Stirling, C., Twomlow, S., & Wollenberg, E. (2018). Urgent action to combat climate change and its impacts (SDG 13): Transforming agriculture and food systems. *Current Opinion in Environmental Sustainability*, *34*, 13–20. doi:10.1016/j.cosust.2018.06.005

Campos, I., Vizinho, A., Coelho, C., Alves, F., Truninger, M., Pereira, C., ... Lopes, G. P. (2016). Participation, scenarios and pathways in long-term planning for climate change adaptation. *Planning Theory & Practice*, *17*(4), 537–556. doi:10.1080/14649357.2016.1215511

Carson, R. (1962). *Silent spring*. Houghton Mifflin.

Carter, J. G., Cavan, G., Connelly, A., Guy, S., Handley, J., & Kazmierczak, A. (2015). Climate change and the city: Building capacity for urban adaptation. *Progress in Planning*, *95*, 1–66. doi:10.1016/j.progress.2013.08.001

Chapple, K. (2015). *Planning sustainable cities and regions: Towards more equitable development*. Routledge.

Clarke, T., McNamara, K. E., Clissold, R., & Nunn, P. D. (2019). Community-based adaptation to climate change: Lessons from Tanna Island, Vanuatu. *Island Studies Journal*, *14*(1), 59–80. doi:10.24043/isj.80

Dagnachew, A., Hof, A., Soest, H. v., & Vuuren, D. v. (2021). *Climate change measures and sustainable development goals*. PBL Netherlands Environmental Assessment Agency.

Das, J. (2019). *Reporting climate change in the Global North and South: Journalism in Australia and Bangladesh*. Routledge. doi:10.4324/9780429402210

Dhar, T. K., & Khirfan, L. (2016). Community-based adaptation through ecological design: Lessons from Negril, Jamaica. *Journal of Urban Design*, *21*(2), 234–255. doi:10.1080/13574809.2015.1133224

Elshater, A., Abusaada, H., & Afifi, S. (2019). What makes livable cities of today alike? Revisiting the criterion of singularity through two case studies. *Cities (London, England)*, *92*, 273–291. doi:10.1016/j.cities.2019 04.008

Elshater, A., Abusaada, H., Alfiky, A., El-Bardisy, N., Elmarakby, E., & Grant, S. (2022). Workers' satisfaction vis-à-vis environmental and socio-morphological aspects for sustainability and decent work. *Sustainability*, *14*(3), 1–34. doi:10.3390u14031699

Fares, A. (Ed.). (2021). *Climate change and extreme events*. Elsevier Science Publishing Co Inc.

Forino, G., Meding, J., & Brewer, G. J. (2015). A conceptual governance framework for climate change adaptation and disaster risk reduction integration. *International Journal of Disaster Risk Science*, *6*(4), 372–384. doi:10.100713753-015-0076-z

Frantzeskaki, N., McPhearson, T., Collier, M. J., Kendal, D., Bulkeley, H., Dumitru, A., Walsh, C., Noble, K., van Wyk, E., Ordóñez, C., Oke, C., & Pintér, L. (2019). Nature-based solutions for urban climate change adaptation: Linking science, policy, and practice communities for evidence-based decision-making. *Bioscience*, *69*(6), 455–466. doi:10.1093/biosci/biz042

Fry, T. (2015). *City futures in the age of a changing climate*. Routledge.

Girardet, H. (2014). *Creating regenerative cities*. Routledge. doi:10.4324/9781315764375

Hongbo, W. (2015). *Global sustainable development report*. United Nations.

Huang-Lachmann, J.-T., & Lovett, C. (2016). How cities prepare for climate change: Comparing Hamburg and Rotterdam. *Cities (London, England)*, *45*, 36–44. doi:10.1016/j.cities.2015.11.001

James, P. (2015). *Urban sustainability in theory and practice: Circles of sustainability*. Routledge.

Kabisch, N., Korn, H., Stadler, J., & Bonn, A. (2017). *Nature-based solutions to climate change in urban areas: Links between science, policy, and practice*. Springer. doi:10.1007/978-3-319-56091-5

Letcher, T. (Ed.). (2021). *The impacts of climate change: A comprehensive study of physical, biophysical, social, and political issues.* Elsevier Science Publishing Co Inc. doi:10.1016/C2019-0-04893-5

Lovejoy, T. E., & Hannah, L. (Eds.). (2019). *Biodiversity and climate change: Transforming the biosphere.* Yale University Press. doi:10.2307/j.ctv8jnzw1

Ma, G., Liu, T., & Shang, S. (2021). Improving the climate adaptability of building green retrofitting in different regions: A weight correction system for Chinese national standard. *Sustainable Cities and Society*, *69*, 102843. Advance online publication. doi:10.1016/j.scs.2021.102843

McHarg, I. (1969). *Design with nature.* John Wiley & Sons, Inc.

Moosmann, L., Siemons, A., Fallasch, F., Schneider, L., Urrutia, C., Wissner, N., & Oppelt, D. (2021). *The COP26 climate change conference, status of climate negotiations and issues at stake.* Luxembourg: Study for the committee on the Environment, Public Health and Food Safety, Policy Department for Economic, Scientific and Quality of Life Policies, European Food Safety.

Okumus, D. E., & Terzib, F. (2021). Evaluating the role of urban fabric on surface urban heat island: The case of Istanbul. *Sustainable Cities and Society*, *73*, 103128. Advance online publication. doi:10.1016/j.scs.2021.103128

Payne, E., & Gordon, B. (2019). The 20th-century architecture of Frank Lloyd Wright. Chicago: Nomination to the World Heritage List by the United States of America (2016) Revised 2019.

Ramyar, R., Ackerman, A., & Johnston, D. M. (2021). Adapting cities for climate change through urban green infrastructure planning. *Cities (London, England)*, *117*, 103316. Advance online publication. doi:10.1016/j.cities.2021.103316

Rauken, T., Mydske, P. K., & Winsvold, M. (2015). Mainstreaming climate change adaptation at the local level. *Local Environment: The International Journal of Justice and Sustainability*, *20*(4), 408–423. doi:10.1080/13549839.2014.880412

Regenesis Group. (2016). *Regenerative development and design: A framework for evolving sustainability.* John Wiley & Sons, Inc.

Roös, P. B. (2021). *Regenerative-adaptive design for sustainable development.* Springer. doi:10.1007/978-3-030-53234-5

Ruskin, J. (1848). *Modern painters* (Vol. 1). Smith, Elder, and Company.

Ruskin, J. (2016). *Nature studies; selections from the writings of John Ruskin*. Wentworth Press.

Secretary-General, U. N., & Ki-moon, B. A. N. (2016). *The sustainable development goals report 2016*. United Nations.

Shearer, H., Coiacetto, E., Dodson, J., & Taygfeld, P. (2016). How the structure of the Australian housing development industry influences climate change adaptation. *Housing Studies*, *31*(7), 809–828. doi:10.1080/02673037.2016.1150430

Singh, C., Soundarya, G. M., Few, R., Kuchimanchi, B., Segnon, A. C., & Morchain, D. (2021). Interrogating 'effectiveness' in climate change adaptation: 11 guiding principles for adaptation research and practice. *Climate and Development*, 1–15. Advance online publication. doi:10.1080/17565529.2021.1964937

Soltani, A., & Sharifi, E. (2019). Understanding and analysing the urban heat island (UHI) effect in Micro-Scale. *International Journal of Social Ecology and Sustainable Development*, *10*(2), 14–28. doi:10.4018/IJSESD.2019040102

Spyridi, D., Vlachokostas, C., Michailidou, A. V., Sioutas, C., & Moussiopoulos, N. (2015). Strategic planning for climate change mitigation and adaptation: The case of Greece. *International Journal of Climate Change Strategies and Management*, *7*(3), 272–289. doi:10.1108/IJCCSM-02-2014-0027

Tabari, H., & Willems, P. (2018). Seasonally varying footprint of climate change on precipitation in the Middle East. *Scientific Reports*, *8*(4435), 1–10. doi:10.103841598-018-22795-8 PMID:29535418

UNEP, UNEP DTU Partnership, & World Adaptation Science Programme (WASP). (2021). *Adaptation Gap Report 2020*. United Nations Environment Programme.

United Nations Framework Convention on Climate Change (UNFCCC). (2016). *The Paris agreement*. United Nations.

Wamsler, C., Wickenberg, B., Hanson, H., Olsson, J. A., Stålhammar, S., Björn, H., ... Zelmerlow, F. (2020). Environmental and climate policy integration: Targeted strategies for overcoming barriers to nature-based solutions and climate change adaptation. *Journal of Cleaner Production*, *247*, 119154. Advance online publication. doi:10.1016/j.jclepro.2019.119154

World Meteorological Organization (WMO). (2021). *State of the climate in Africa 2020*. Geneva: WMO. Retrieved from https://library.wmo.int/index.php?lvl=notice_display&id=21973#.YckILRNBz6O

Wright, F. L. (2012). *Drawings and plans of Frank Lloyd Wright: The early period (1893-1909)*. Dover Publications.

## KEY TERMS AND DEFINITIONS

**Adaptive Capacity:** Focuses on climate change and adjusts a system to external factors from its environment. Increasing adaptive capacity is a realistic way to deal with climate change, including unpredictability and extremes. When the adaptive capability is improved, the vulnerabilities are reduced.

**Regenerative City:** Concept stresses the importance of moving beyond a narrow definition of sustainability and embracing a holistic approach to urban planning. This approach stresses that cities need to do more than keep up with the natural resources that they use to live and grow.

## ENDNOTES

1 In April 2016, 175 Member States signed the historic Paris Agreement (UN Secretary-General BAN Ki-moon, 2016, p. 9), which intends to enhance the global response to climate change by limiting global temperature rises below 2° Cover pre-industrial levels this century (United Nations Framework Convention on Climate Change (UNFCCC), 2016).

2 Goal 13 mandates swift action not only to address climate change and its consequences but also to strengthen resilience in the face of climate-related hazards and natural disasters (United Nations Framework Convention on Climate Change (UNFCCC), 2016, p. 36).

3 This report focuses on the challenges of financing adaptation initiatives in developing countries that safeguard against climate impacts such as droughts, floods, and sea-level rise (UNEP, UNEP DTU Partnership, World Adaptation Science Programme (WASP), 2021). The impact of rising temperatures within cities, particularly in metropolitan areas, was not mentioned in this study. One of the key motivations for drafting this chapter is to shed light on this gap.

4 This conference focused on promoting and implementing the Paris Agreement as well as the international framework for addressing climate change, the parties and stakeholders participating, the status of the negotiations, and recent developments that have influenced the international process (Moosmann, et al., 2021).

5 Another concern in underdeveloped countries that has been exacerbated by climate change is the lack of shelter for stray dogs and free-roaming cats in hot and cold regions (Abusaada & Elshater, 2021a). This aspect is overlooked as many people feel that the poor people are the most caring than stary dogs or free-roaming cats.

# APPENDIX 1

*Table 1. Fourteen books climate about climate change, regenerative cities, sustainability, sustainable development goals between 2015-2021*

| No. | Date | Publisher | Title | Keywords | References |
|---|---|---|---|---|---|
| 1. | 2021 | Elsevier Science Publishing Co Inc | The impacts of climate change: A comprehensive study of physical, biophysical, social, and political issues | climate change | (Letcher, 2021) |
| 2. | 2021 | PBL Netherlands Environmental Assessment Agency | Climate change measures and sustainable development goals | climate change, sustainable development goals | (Dagnachew, Hof, Soest, & Vuuren, 2021) |
| 3. | 2021 | Springer | Regenerative-adaptive design for sustainable development | regenerative-adaptive design, sustainable development | (Roös, 2021) |
| 4. | 2021 | Elsevier Science Publishing Co Inc | Climate change and extreme events | climate change | (Fares, 2021) |
| 5. | 2019 | Routledge | Reporting climate change in the Global North and South: Journalism in Australia and Bangladesh | climate change | (Das, 2019) |
| 6. | 2019 | Yale University Press | Biodiversity and climate change: Transforming the biosphere | biodiversity, climate change, biosphere | (Lovejoy & Hannah, 2019) |
| 7. | 2018 | Routledge | The Routledge companion to planning in the Global South | Global South | (Bhan, Srinivas, & Watson, 2018) |
| 8. | 2017 | Springer | Nature-based solutions to climate change in urban areas: Links between science, policy, and practice | climate change, nature-based solutions | (Kabisch, Korn, Stadler, & Bonn, 2017) |
| 9. | 2016 | John Wiley & Sons, Inc | Regenerative development and design: A framework for evolving sustainability | regenerative, sustainability | (Regenesis Group, 2016) |
| 10. | 2015 | Routledge | Creating regenerative cities | regenerative cities | (Girardet, 2015) |
| 11. | 2015 | Routledge | City futures in the age of a changing climate | climate change | (Fry, 2015) |
| 12. | 2015 | Routledge | Sustainability in architecture and urban design | sustainability, urban design | (Bovill, 2015) |
| 13. | 2015 | Routledge | Sustainable urban planning | sustainable urban planning | (James, 2015) |
| 14. | 2015 | Routledge | Planning sustainable cities and regions | planning | (Chapple, 2015) |

*Table 2. Eighteen articles in sixteen journals include adaptation, climate adaptability, climate change, climate change adaptation, and urban heat island as a keyword between 2015-2021*

| No. | Date | Journals | Title | Keywords in title and abstract | Used keywords | References |
|---|---|---|---|---|---|---|
| 1. | 2021 | Sustainable Cities and Society | Evaluating the role of urban fabric on surface urban heat island: The case of Istanbul | urban heat island, urban fabric | urban heat island | (Okumus & Terzib, 2021) |
| 2. | 2021 | Sustainable Cities and Society | Improving the climate adaptability of building green retrofitting in different regions: A weight correction system for Chinese national standard | building green, climate adaptability, | climate | (Ma, Liu, & Shang, 2021) |
| 3. | 2021 | Cities | Adapting cities for climate change through urban green infrastructure planning | climate change, urban green infrastructure planning | adaptation, climate change | (Ramyar, Ackerman, & Johnston, 2021) |
| 4. | 2021 | Climate and Development | Interrogating 'effectiveness' in climate change adaptation: 11 guiding principles for adaptation research and practice | adaptation climate change | adaptation, climate change | (Singh, et al., 2021) |
| 5. | 2020 | Journal of Cleaner Production | Environmental and climate policy integration: Targeted strategies for overcoming barriers to nature-based solutions and climate change adaptation | adaptation, climate change adaptation, climate policy integration, strategies, nature-based solutions | adaptation, climate change | (Wamsler, et al., 2020) |
| 6. | 2019 | Island Studies Journal | Community-based adaptation to climate change: lessons from Tanna Island, Vanuatu | adaptation, climate change, community-based adaptation | adaptation, climate change | (Clarke, McNamara, Clissold, & Nunn, 2019) |
| 7. | 2019 | BioScience | Nature-based solutions for urban climate change adaptation: Linking science, policy, and pactice communities for evidence-based decision-making | adaptation, climate change, Nature-based solutions, urban adaptation | adaptation climate change | (Frantzeskaki, et al., 2019) |
| 8. | 2019 | International Journal of Social Ecology and Sustainable Development (IJSESD) | Understanding and analysing the urban heat island (UHI) effect in Micro-Scale | urban heat island, Micro-Scale | urban heat island climate change | (Soltani & Sharifi, 2019) |
| 9. | 2018 | Current Opinion in Environmental Sustainability | Urgent action to combat climate change and its impacts (SDG 13): transforming agriculture and food systems. | climate change, combat, | climate change | (MCampbell, et al., 2018) |
| 10. | 2016 | Journal of Planning Education and Research | Equity impacts of urban land use planning for climate adaptation: Critical perspectives from the Global North and South | climate adaptation | adaptation climate change | (Anguelovski et al., 2016) |
| 11. | 2016 | Planning Theory & Practice | Participation, scenarios and pathways in long-term planning for climate change adaptation | adaptation, climate change, | adaptation climate change | (Campos et al., 2016) |
| 12. | 2016 | Journal of Urban Design | Community-based adaptation through ecological design: lessons from Negril, Jamaica | community-based adaptation, ecological design | adaptation | (Dhar & Khirfan 2016) |
| 13. | 2016 | Cities | How cities prepare for climate change: Comparing Hamburg and Rotterdam | climate change | climate climate change | (Huang-Lachmann and C.Lovett, 2016) |
| 14. | 2016 | Housing Studies | How the structure of the Australian housing development industry influences climate change adaptation | adaptation, climate change, climate change adaptation | adaptation, climate change | (Shearer, Coiacetto, Dodson, and Taygfeld, 2016) |
| 15. | 2015 | Progress in Planning | Climate change and the city: Building capacity for urban adaptation | adaptation, climate change, urban adaptation | adaptation, climate change | (Carter et al., 2015) |
| 16. | 2015 | International Journal of Disaster Risk Science | A conceptual governance framework for climate change adaptation and disaster risk reduction integration | adaptation, climate change | adaptation, climate change | (Forino, Meding, and Brewer, 2015) |
| 17. | 2015 | Local Environment: The International Journal of Justice and Sustainability | Mainstreaming climate change adaptation at the local level | adaptation, climate change, local level | adaptation, climate change | (Rauken, Mydske, and Winsvold, 2015) |
| 18. | 2015 | International Journal of Climate Change Strategies and Management | Strategic planning for climate change mitigation and adaptation: The case of Greece | adaptation, climate change, mitigation, strategic planning | adaptation, climate change | (Spyridi, Vlachokostas, Michailidou, Sioutas, and Moussiopoulos, 2015) |

# Chapter 2
# Strides Towards an Eco-Friendly and Regenerative City Future

**Lina Fayed**
*German International University, Egypt*

**Ahmed F. Amin**
*American University in Cairo, Egypt*

## ABSTRACT

*Urbanisation affects the environment and ecosystems in many ways. For example, cities are continuous, resulting in their footprint being extended. Rapid urbanisation is driving cities to become significantly warmer than the surrounding countryside as a result of this rapid urbanisation and the unsustainable attitude of their citizens toward the city. The Sustainable Development Goals (SDGs) were adopted by countries worldwide a few years ago. This will enable challenges. In light of this, cities need to adopt sustainable urbanization, allowing citizens to participate in participatory, integrated, sustainable city planning and city management to enhance the quality of life. By implementing the SDGs, cities can help reverse the damage done to their ecosystems. As part of its 2052 National Vision, Egypt has also implemented the SDGs and achieved targeted improvements for its citizens. This chapter concludes with comprehensive initiatives in the context of implementing the SDGs.*

## INTRODUCTION

According to the United Nations (UN), the latest Urban Agenda aims to reconsider

DOI: 10.4018/978-1-6684-2462-9.ch002

the methodology used for planning, designing, financing, developing, and managing cities to ensure that cities can accommodate almost 60% of the world's population by 2030. In this timeframe, the current rate of urbanisation in cities will inevitably increase the urban land cover footprint three times and accommodate more than 1.024 billion people (Seto, Güneralp, & Hutyra, 2012). Cities are removing natural green lands to replace them with paved surfaces, buildings, and other surfaces that absorb and retain heat, producing urban heat islands. The urban heat island effect leads to increased energy costs; increased air pollution increases heat-related mortality (Elmarakby, Khalifa, Elshater, & Afifi, 2022).

2021 has witnessed climatic and environmental changes, starting with the high heat waves in countries worldwide, basically across the Globe. Additionally, the United Kingdom, Austria, Belgium, Croatia, Germany, Luxembourg, the Netherlands, Switzerland, Italy, Turkey, and Japan have experienced out-of-season flooding. The rate of spread and the size of forest fires have also been unprecedented in this year's fire season in countries such as the United States of America, Italy, Greece, Turkey, and Algeria.

According to the UN report on these catastrophic current events, this human-induced climatic change results in weather and climate extremes phenomena worldwide. The report highlights that global human influence and current practices have warmed the climate unprecedentedly in the past 2,000 years (UN Climate Change News, 2021).

This chapter aims to influence cities to take responsibility for their role in the environmental degradation that the world has reached today. As such, cities need to take action to prevent further degradation and reverse the damage done to the environment while enhancing the ecosystems that are currently being degraded. However, the solution to that is clear. Cities and countries should adopt net-zero emissions policies and improve national planning policies to reach inclusive and green economies, prosperity, cleaner air, and better health for their citizens (UN Climate Change News, 2021).

## ENVIRONMENTAL NEED FOR SUSTAINABILITY

The world's population is constantly increasing, and cities are continuously growing, so morden cities should become more inclusive to accommodate everybody. Furthermore, the world needs to build modern and sustainable cities that consider their ecosystems. For cities' citizens to survive and thrive, cities need a shift to new urban planning strategies that produce safe, affordable, and resilient cities with green and culturally inspirational surroundings (United Nations, 2015).

Transforming cities to become safe and sustainable requires ensuring citizens' accessibility to safe and affordable housing, encouraging investment in public transportation, producing green public spaces, and enhancing urban planning and management in a participatory and inclusive (Eurostat, 2020). Cities and towns mustn't function separately from rural areas. They should mainly look towards their rural surroundings and depend on them to sustain their needs. This relationship between rural-urban division in modern cities should be broken down, and a more sustainable food-system approach should be promoted (Nuruzzaman, 2015). If cities succeeded in encouraging urban innovation in thoughtful, ground-breaking ways, cities would be able to assure that "no one and no place" is left behind.

Cities should follow urban strategies that would reduce the urban heat island effect. Also, cities should implement strategies that enhance their ecological impact through a comprehensive vision. This new approach offers opportunities for all citizens and access to essential services, energy, housing, transportation, and green public spaces while reducing resource consumption and the accompanied negative environmental impact.

Furthermore, cities should adopt other environmentally friendly strategies. Even if public spaces in towns and urban areas are limited, the cities should seek to integrate small green infrastructure practices into any grassy or barren regions available in the city and the city's vacant lots and street rights-of-way as a central urban planning strategy. Thus, the city can use urban strategies to minimise urban heat, reuse stormwater, and other concerns, inevitably leading to healthier environments and enhanced citizens' lifestyles. As urban areas continue to expand, green spaces are disappearing. More than just for aesthetic appeal, public spaces and green areas are essential for improving air quality, mitigating urban temperatures, encouraging physical activity and improving overall health, and reducing the urban heat island effect.

Since 2015, counties worldwide have adopted a comprehensive approach to address their ecosystems' challenges. These countries, and consequently their cities, have committed to achieving the SDGs to provide a better and more sustainable future for all (United Nations, 2016). In general, for Cities to reach this vision, they have to address the global challenges that face their citizens, such as poverty, inequality, climate change, environmental degradation, peace, and justice (United Nations, 2019).

The Sustainable Development Goals (SDGs) are a collection of 17 interconnected goals designed to be a "blueprint that achieves a better and more sustainable future for all citizens" (United Nations, 2015). The SDGs have 17 goals that collectively embark on the path toward cities' sustainable development (Figure 1). The 2030 Agenda of the Sustainable Development Goals highlights vital links between development, the environment, citizens' well-being, and the complete possession of a wide range of human rights (United Nations Environment Programme, 2016).

*Figure 1. The 17 sustainable development goals*

Today, cities and countries worldwide are committed to improving a range of environmental sectors, enhancing public health and significant economic, financial, and social gains in the past decades. Since the 2000s, the world has achieved the Millennium Development Goal (MDGs) target of reducing by half the percentage of citizens with no access to improved water sources. Also, cities succeeded in the phase-out of nearly 100 ozone-depleting substances means that almost 2 million cases of skin cancer per year will be prevented by 2030 thanks to the healing of the ozone layer. However, many challenges remained. The cities' ecosystems that sustain their citizens are estimated to be responsible for 23% of global deaths, which calls for a more comprehensive measure (United Nations Environment Programme, 2016). In 2012 alone, almost 12.6 million deaths worldwide were related to environmental health implications.

Accordingly, a successor to the MDGs was needed for 2015–2030. There was a demand for SDGs to be developed, as they were intended to be a more comprehensive approach that understood the learned lessons of the MDGs and deals with the constantly growing challenges of modern life. The SDGs being much broader than the MDGs, they are more capable of impacting the cities' ecosystems (Asi, 2021). Cities' craving for sustainable development has led to various development objectives and the development of the needed tools and measures for structuring and conducting sustainable development policy analyses (Ogujiubm & Ehigiamusoe, 2013).

All countries face particular challenges while pursuing sustainable development (UNnited Nations, 2017). The least developed and most fragile states would currently have more challenges to overcome (Asi, 2021). As a result, economically developed countries that can meet some of the MDGs more quickly now have to be judged on a broader scale.

Generally, sustainable development is about meeting the needs of the present without jeopardising the chances of future generations meeting their own needs. The SDGs take a global approach to city development rather than the standard indicators that strictly measure the baseline achievements. As for the regional context, the Gulf Arab states rank highly, indicating high levels of development. Yet a few countries are on track to achieve most of the SDGs. Another look at the Middle East and North Africa region's performance in the SDGs shows that there are a lot of problems with meeting them (Ogujiubm & Ehigiamusoe, 2013).

This chapter mainly addresses SDG 11, the Sustainable Cities and Communities goal that aims to accommodate everyone, so modern, sustainable cities are needed. New, intelligent urban planning is required to make cities safe, affordable, resilient, and green so that all city residents can live and thrive (The Global Goals, 2015).

By 2030, SDG 11, "Sustainable Cities and Communities", aims to realise many cities' objectives. They seek to ensure access by city inhabitants to adequate, safe, and affordable housing and to upgrade slums. In addition to improving accessibility to sustainable transport systems and enhancing road safety, they expanded public transport, considering the needs of vulnerable categories. Also, cities should enhance numerous aspects of their context, such as sustainable urbanisation, community participation, city-wide sustainable planning, and sustainable urban environmental management. Furthermore, cities should make more efforts to protect and safeguard their citizens' cultural and natural heritage (United Nations, 2016).

Additionally, SDG 11, "Sustainable Cities and Communities", enables cities to reduce the number of deaths and substantially decrease the direct economic losses relative to the global gross domestic product caused by disasters, including water-related disasters, and protect the poorest citizens in vulnerable situations. Another target is to offer all city citizens access to green and public spaces that are safe, inclusive, and accessible, specifically for women and children, the elderly, and the disabled (United Nations, 2015).

Moreover, SDG 11, "*Sustainable Cities and Communities*", has key targets that tackle the cities' modern challenges. These key targets aim to reduce the negative environmental impact of cities by giving special attention to air quality and municipal and other waste management (United Nations, 2015). By 2030 cities should have acquired positive economic, social, and environmental links between urban and rural areas through boosting national and regional development planning. According to the "*Sustainable Cities and Communities*" goal, even more cities should adopt and implement integrated policies towards resources efficiency, mitigation, and adaptation to climate change, resilience to disasters, and develop a holistic disaster risk management at all levels. The SDGs encourage the less developed countries to have sustainable and resilient buildings that utilise local materials through financial and technical aspects (Figure 2).

The environment is essential in SDG 11, which supports cities to become low-emission, resource-efficient, and resilient to reduce cities' environmental impact on their ecosystems and their negative environmental impacts (Takase, 2018). The rapid urbanisation of modern cities has a deteriorating impact on their ecosystems, such as climatic change, global warming, heat island effect, etc. Focusing more on the heat island effect phenomenon, it happens when cities replace the natural land cover with vast concentrations of pavements, buildings, and other surfaces that absorb and maintain heat (U.S. Environmental Protection Agency, 2016). This urban heat island occurs when an urban area or metropolitan area is much warmer than its rural surroundings due to human actions (Hideki Takebayashi, 2020).

## IMPACTS OF URBAN HEAT ISLANDS ON MODERN CITIES

The urban heat island effect is a micro-climatic phenomenon, which differs with city size. The phenomena reflect the differences in air temperature between urban and suburban areas compared to rural surroundings (Filho, et al., 2021; Elmarakby,

*Figure 2. The SDG 11 key targets, Source: (The Global Goals, 2015)*

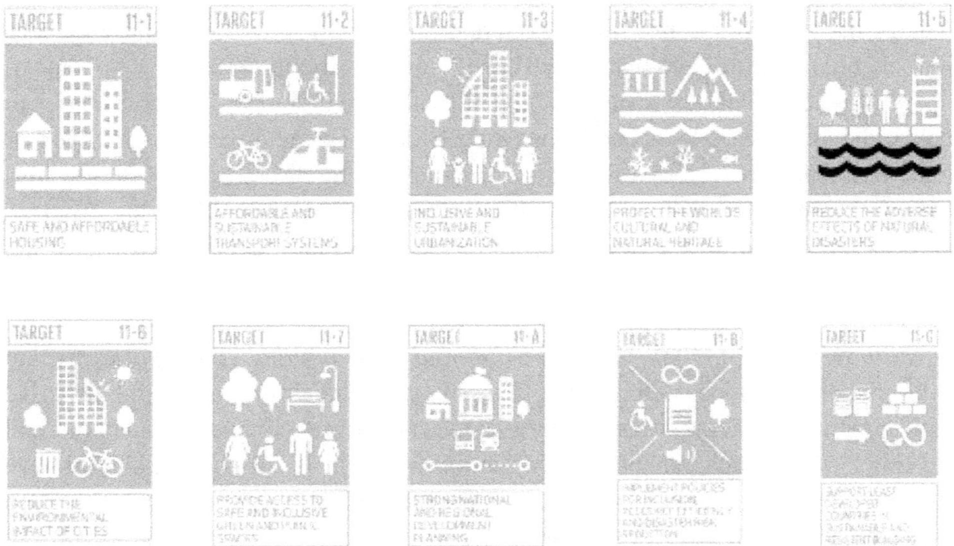


Khalifa, Elshater, & Afifi, 2022). Furthermore, the continuous cities' urban growth has the most significant impact on the UHI effect as Natural landscapes consist of different porous surfaces, such as greenery and soil surfaces. On the other hand, cities are full of impervious surfaces such as bitumen and concrete. Unlike greenery and other natural surfaces, modern building materials absorb heat rather than reflect it, leading to raised temperatures (Appleton, 2020). This phenomenon also raises energy costs, air pollution levels, and heat-related disease and deaths within the city compared to the rural areas (Figure 3).

As a result of cities' urban development, their accompanied rapid population growth and urban expansion, along with the dense urban infrastructure, have enormously affected the quality of life, negatively influenced the urban environment, and damaged the cities' ecosystems (El-Magd, Ismail, & Zanaty, 2016). The causes of the phenomena are mainly rapid urbanisation, increasing density of buildings, and the changing land surface in cities and transformed urban environment (Mirzaei & Haghighat, 2010). Furthermore, reductions in evapotranspiration, convection, and increases in heat storage, net radiation, and anthropogenic heat expand the urban heat islands. The above factors produced cities with diminished greenery and overuse of impermeable surfaces (Mohajerani, Bakaric, & Jeffrey-Bailey, 2017).

Other influencing factors are heat pockets between high buildings, waste heat resulting from industry and transportation, and the citizens' energy consumption

*Figure 3. The urban heat island phenomena in urban areas vs rural areas. Source: (Adhikari, 2018)*

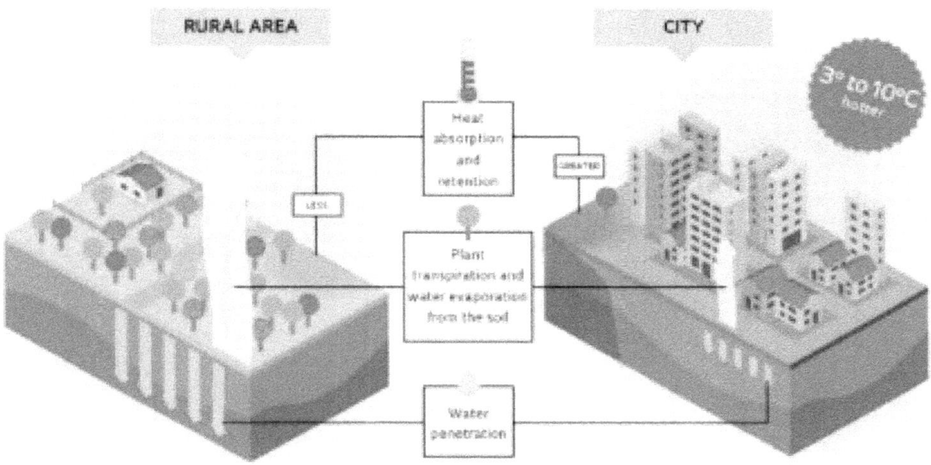


and unsustainable lifestyle, in addition to the problem of the never-ending cycle of excessive use of air conditioners. While cities become warmer, citizens will need more air conditioning to keep on functioning, leading to more wasted heat that increases the temperature in a cyclical pattern (Appleton, 2020) and as shown in Figure 4. Therefore, cities should utilise certain urban materials and geometries to relieve heat, slow the winds, and reduce air pollution and energy consumption. Cities are adopting the SDGs to address their multiple environmental, social, and economic challenges.

Cities should implement more comprehensive strategies to reduce the urban heat island effects through trees, green roofs, and vegetation, which lead to shading building surfaces, deflecting the radiation from the sun, and releasing moisture into the atmosphere (U.S. Environmental Protection Agency, 2016). Also, green roofs are an effective heat island reduction strategy, as it provides both direct and indirect cooling effects. Additionally, green roofs improve air quality by minimising the heat island effect and pollution absorption. Furthermore, cities should build green infrastructure and integrate it into the regular street upgrade and city upgrading projects to ensure continuous investment in heat-reducing implementation strategies.

The urban heat island effect has multiple implications on the city and its citizens, such as the spread of infectious diseases, extensive heat causing dehydration, and an increase in the number of morbidity and fatality, especially for those working in outdoor conditions. Other environmental impacts are changes in weather and climatic conditions, degraded water quality, and increased emission of air pollutants

*Figure 4. Factors contributing to the urban heat island effect. Source: (Tan, Kolokotsa, & Takebayashi, 2021)*

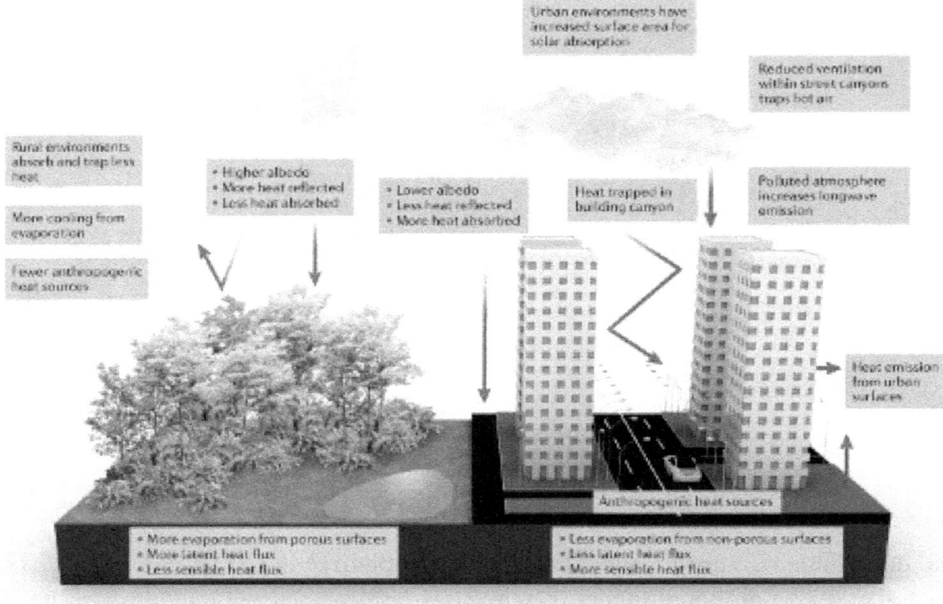


and greenhouse gases, which naturally affect the city's ecosystems and biocapacity (Adhikari, 2018).

The SDGs call for adopting a sustainable urban planning strategy by strategically including urban and peri-urban forests in the city's national vision and promoting a productive rural-urban relationship. Furthermore, cities need Comprehensive solutions that would go hand-in-hand with further sustainable development goals such as reducing poverty, improving quality of life, and promoting sustainable economic development. A 'green agenda' may thus be a vital part of a holistic, city-led strategy toward enhanced economic, social and environmental sustainability (Filho, et al., 2021).

Nowadays, cities and metropolitan areas are growing at exceptional rates (Figure 5). Thus, countries worldwide seek to achieve the SDG 11 "*Sustainable Cities and Communities*" and its targets to increase the cities' urban resilience to climate change impacts and disasters. And so, countries that include SDG 11 in their national visions consider cities as a socio-economic force, focusing on challenges related to cities' rapid urbanisation, segregation of urban settlements, and increased air pollution (Devisscher, et al., 2019). Therefore, countries should implement the SDGs comprehensively while giving extra attention to the SDG11, its implementation

*Figure 5. Green spaces cooling effect, Source: (Griloa, et al., 2020)*

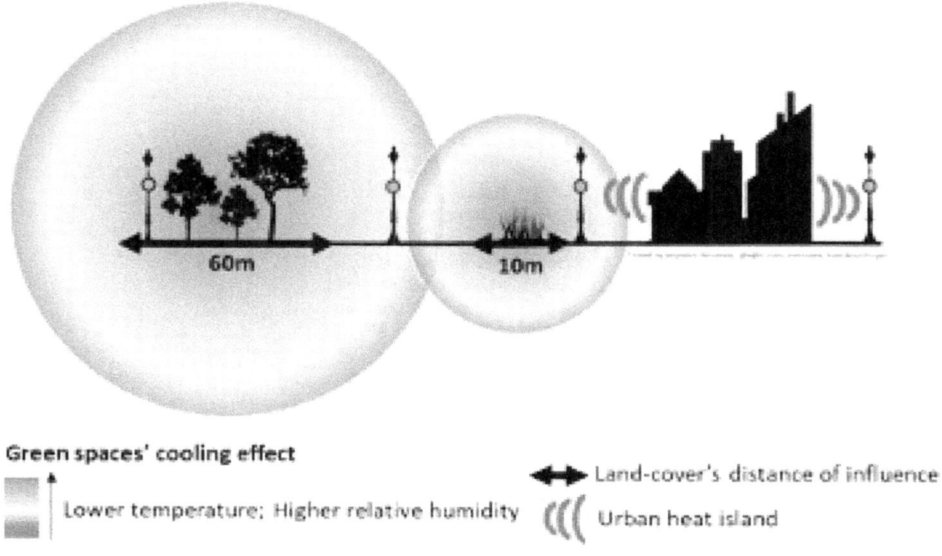


strategies, and key targets that address the cities' currently deteriorated environment, climatic change that resulted in urban heat island effects.

Moreover, SDG 11, "*Sustainable Cities and Communities",* promote public and green spaces, which offer opportunities to advance the well-being of citizens in cities. The SDG 11 key targets will help reduce the urban heat island effect as sustainable transportation, urbanism, and green spaces will bring down city temperatures and contribute to a more energy-efficient city. Increasing green spaces in the town will help reduce this phenomenon through the cooling effect of green spaces, providing more shade, and retaining air pollutants (Farshid Aram, 2019).

## BACKGROUND OF SDGS IMPLEMENTATION EFFORTS IN EGYPT

Today, multiple tensions are playing out throughout the Egyptian regional context - specifically the Arab region - and specifically with the disruption caused by the COVID-19 pandemic. The United Nations recognises three connected dimensions to sustainable development, which are economic, social, and environmental. Sustainable development in Africa includes various aspects such as poverty, increased population, rapid urbanisation, and the environmental impact of extractive industries (Asi, 2021).

*Figure 6. The Sustainable Development Strategy (SDS). Source: (The Ministry Of Planning and Follow-up and Administrative Reform, 2016)*

Since 2015, Egypt has committed to the Sustainable Development Goals as it focuses on goals aligned with the local Sustainable Development Strategy (SDS). SDGs are considered a transformational vision that can achieve the aspiration of cities' citizens. Accordingly, the Egyptian government seeks to achieve balanced development management of land and resources to reflect the needs of citizens and enhance the quality of life offered to citizens (Figure 6).

In Figure 7, the Egyptian national SDGs strategy aims to develop a comprehensive, long-term political, economic, social, and environmentally aware sustainability

*Figure 7. National long-term targets*

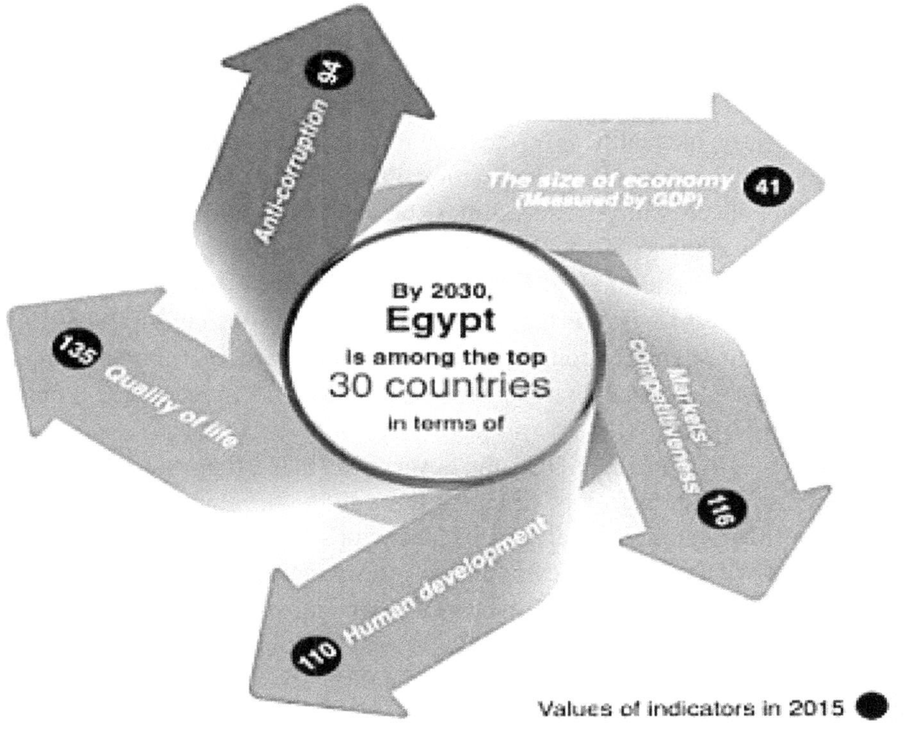


vision to achieve its long-term national targets. This long-term vision is the base for the short- and medium-term development plans at the different levels, such as the national, local, and sectoral levels, to reflect the expectations of Egyptians. Additionally, the strategy seeks to involve all the stakeholders, such as the private sector and the civil society, to achieve the SDG's vision and goals. It allows the parliament to monitor the implementation of the strategy's objectives, targets, programs, and projects. As for the regional and global levels, the approach aligns its goals with the 2030 plan and the African Union's Sustainable Development Strategy for 2063 (The Ministry of International Cooperation, 2016). Localising the SDGs through the SDS is adopted by multiple ministries and entities, which was used to develop the Egyptian medium- and long-term strategies (The Ministry of Planning, Monitoring and Administrative Reform, 2018).

Furthermore, sustainable development is connected to the carrying capacity of cities' ecosystems. Also, sustainable development and urbanisation are considered universal challenges on their own (Fayed, Elshater, & Rashed, 2020). The rapid urbanisation of Egyptian cities with no environmental awareness means that in almost every urban centre, a substantial proportion of the population is at risk from natural and human-induced ecological hazards (Ogujiubm & Ehigiamusoe, 2013).

Accordingly, Egypt is actively preparing and negotiating for implementing this new paradigm shift. The Egyptian government is committed to effectively implementing this historical agreement within mutually-beneficial cooperation, aiming to end all forms of poverty, inequalities and tackle climatic change while assuring that no one is left behind (The Ministry of International Cooperation, 2016).

As Egypt successfully achieved most of the Millennium Development Goals (MDGs) from 2000 to 2015, these goals ranged from halving extreme poverty to reverting the spread of HIV/AIDS and acquiring universal primary education. This is a massive success for the country, taking into consideration that Egypt is already the most populated country in the Arab region, where its population growth is expected to be around 24%, from 93.8 million in 2015 to reach 122.6 million in 2030 (David K. Bohl, 2018).

While implementing the SDGs, the government is learning from the past; Egypt is committing to a transformative vision through the development of the national Sustainable Development Strategy (SDS): Egypt's Vision 2030, depending on the embodiment of the new strategies, addressing welfare and prosperity as primary goals, that would be achieved by sustainable development, social justice, and balancing geographical and social growth (The Ministry of International Cooperation, 2016). With cities following the Egyptian SDGs and the 2030 Vision state, Egypt will achieve a competitive, well-balanced, diversified economy. These new strategies will offer citizens justice, social integration, and participation and provide cities with a balanced and enhanced ecosystem. This vision benefits Egypt's strategic location and human capabilities to realise sustainable development and a better life for Egyptians (The Ministry of International Cooperation, 2016).

## EGYPT'S PROGRESS TOWARDS GOALS AND TARGETS

This part of the chapter briefly shows Egypt's progress toward achieving the SDGs. Then it will discuss more thrilly the SDG 11 "*Sustainable Cities and Communities*". According to the United Nations ESCWA report "*Reporting Platforms for the Sustainable Development Goals: Assessment*," Egypt scored the highest - in comparison to the Arab region - in implementing the Sustainable Development Goals by 50% (Figure 8).

The above-mentioned report stated the key SDGs, which achieved remarkable progress in agriculture, communications, and technology. In the agriculture SDGs, Egypt's succeeded in increasing the percentage of agricultural area under productive and sustainable agriculture from 3.7% in 2013 to 3.8% in 2015. Also, there was

*Figure 8. Countries SDGs Achievement Score, Source: (Egyptian Streets, 2020)*

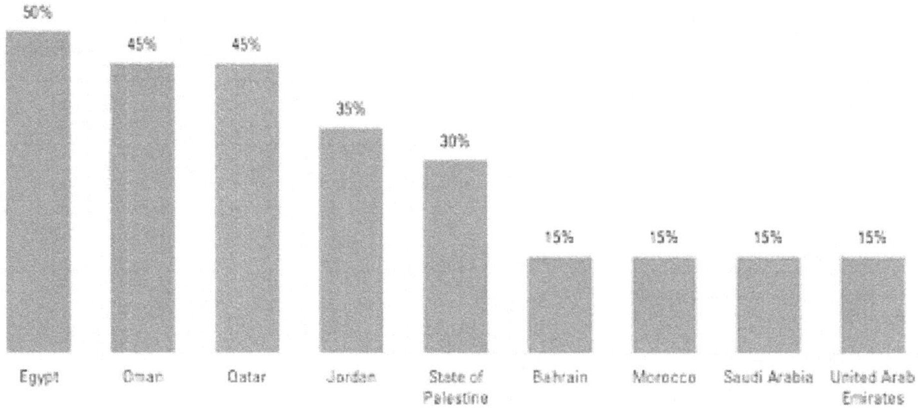


another progress in communications and technology, where the percentage of citizens accessing the Internet increased from 24.7% in 2014 to 29.2% in 2015 (Figure 9).

*Figure 9. Projects implemented among the SDGs, Source: (Egyptian Streets, 2020)*

As for the SDG 11 "*Sustainable Cities and Communities*", as mentioned above, Egypt has one of the highest population growth rates in the region. This high population density led to numerous challenges, such as the deterioration of the quality of urban life and damaging the environment. Consequently, Egypt suffers from inadequate urban expansion- the whole population lives on merely 6.8% of Egypt's total area - accompanied by unbalanced regional development across Egypt.

*Table 1. Indicators of the "Sustainable Cities and Communities" goal. Source: (The Ministry of Planning, Monitoring and Administrative Reform, 2018)*

| Indicators | SDG target | Types | Value | 2030 Target | Changes |
|---|---|---|---|---|---|
| Total population density (population / km$^2$) | 11.1 | 0 | 96.6 …… 2018 | N/A | – |
| The inhabited area as a percentage of total area (%) | 11.1 | 0 | 6.8 …… 2017 | N/A | * |
| Percentage of households connected to the public water network (%) | 11.1 | 2 | 96.9 …… 2017 | 100 | + |
| Percentage of households connected to the public sanitation network (%) | 11.1 | 2 | 55.9 …… 2017 | 100 | + |
| The proportion of slum areas to a total urban size (%)- unsafe | 11.1 | 2 | 0.7 …… 2017 | 0 | × |
| The proportion of slum areas to a total urban size (%)- unplanned | 11.1 | 2 | 37.5 …… 2017 | 25 | × |
| The proportion of new cities with waste management systems (%) | 11.6 | 2 | 60 …… 2017 | 100 | × |
| Per capita green landscapes in cities (m$^2$/capita) | 11.7 | 2 | 0.85 …… 2015 | 3 | × |

+ is positive change; – is negative change; * is no change; × is positive change but data is not available

This unbalanced development led to the migration of residents from rural to urban areas (particularly to major cities such as Cairo and Alexandria) to pursue better job opportunities, health, and education. Another consequence of the major cities' large population is the spread of slums, traffic congestion, pollution, building on agricultural land, and poor economic and social conditions of rural-to-urban migrants (The Ministry of Planning, Monitoring and Administrative Reform, 2018).

Realising the need for integrated and sustainable urban planning strategies, the SDS is built on the notion of balancing regional development with local development goals. The national urban development system is aligned with SDG 11, which aims to increase inhabited areas, enhance the quality of urban life, and make the most out of Egypt's strategic location. Egypt's success in achieving the SDG 11 objectives varied among the "Sustainable Cities and Communities" goal (Table 1).

Egypt is developing new cities all over the country as an essential part of its urban strategies. These new cities would increase the percentage of inhabited areas and reduce high population densities. Currently, around fifteen new cities are being built. These cities are geographically distributed due to the government's policies toward a more balanced regional development. These new cities' generations are designed to be green, and energy sustainable, as the per capita green landscape is planned to increase consistently from 0.85 m2 in 2015 to 3 m2 in 2030 (CAPMAS, 2017).

*Figure 10. Sustainable Cities Criteria. Source: (The Ministry of Planning, Monitoring and Administrative Reform, 2018)*

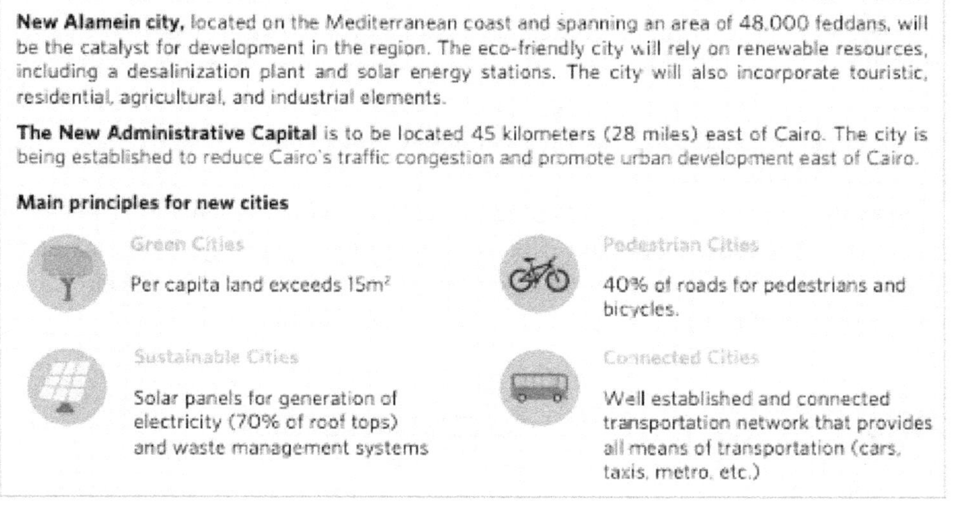


New cities such as new Alamein and the New Administrative Capital are planned to be sustainable green cities that consider their ecological footprint, focusing on sustainable planning and ecofriendly practices (Figure 10).

Cairo has a reported residents density of 50,259 people per square kilometre. And so it is considered one of the top five most dense cities worldwide. With the continuous growth in population and the rapid rural-to-urban migration, Cairo has also been experiencing increased growth in the population of slums resulting in more development of informal settlements in the city and the surrounding districts (The Ministry of Planning, Monitoring and Administrative Reform, 2018).

Furthermore, developing slums and informal settlements is a milestone in achieving SDG11 as well as it has been considered one of the major priorities. Several efforts have been implemented to provide safe and adequate living conditions for the residents of these informal settlements. The "Informal Settlements Development Facility (ISDF)" was founded to support local government in providing safe housing, prioritising the eradication and development of informal settlements. The scope of the ISDF ensures that its development system would realise the SDGs, offering more than a safe and healthy home but also improving the quality of life and putting an end to extreme poverty for people living in these areas (The Ministry of International Cooperation, 2016).

Egypt's arid climate and continuous rapid urbanisation were magnified by urban heat island impacts on Egyptian cities (Alchapar & Cantón, 2016). The rapid urban growth of new cities in Egypt and the expansion growth of the Greater Cairo Region were accompanied by an increase in the heat island effect. This rapid urbanisation is considered the main contributor to global warming and climatic change in megacities (Elshater, et al., 2022). In addition, Cairo is one of the densest cities worldwide, which leads to significant temperature variation compared to the rural in the Greater Cairo Region (El-Magd, Ismail, & Zanaty, 2016). This is urgent to address the current environmental challenges to restore the damage to the city's ecosystem in alignment with the SDGs and Egypt's SDSs.

Nevertheless, the Egyptian government's commitment to implementing the SDGs can help minimise this heat island effect. As shown above the projects implemented address the agricultural land and the greenification of the city, which can help reduce the side effects of the urban heat island phenomena (Shishegar, 2014).

## CONCLUSION: A PATHWAY TO FUTURE CITIES

As a result of the world's current environmental and climatic frightening challenges, it became urgent for cities to integrate sustainable development goals (SDGs) as a critical aspect of their sustainable development policies. Additionally, it is essential to align the sustainable development policies with the various efforts of the government to achieve maximum coherence and efficiency in implementation. It is also necessary to enhance the coherence of international support to national sustainable urban development plans as part of an overall national sustainable development planning.

The implementation of the SDGs still faces some challenges, such as the participatory framework, the management and coordination process, and economic and social factors that influence the rate of accomplishment. For example, the participation and role of different stakeholders involved in the process remain confusing. Further stakeholder analysis should be done to identify other entities affecting the achievement of the SDGs, stakeholders such as the Government, civil society, and the private sector. These stakeholders can choose based on their influence or interest and whether they have a common interest.

Countries and cities worldwide have considered numerous active and passive strategies to address the UHI effect. Furthermore, the SDG 11 "Sustainable Cities and Communities" implementation still faces innumerable challenges despite global efforts, such as the lack of consistent data collection and collective methodology, leading to inconsistencies, gaps, and focusing on specific targets more than others. The implementation initiatives should reflect cities' crucial needs when prioritising specific SDG 11 targets before others. Specifically, more attention should be given to climate change and global warming due to the natural and environmental crises this year. People should also pay attention to the increasingly alarming climate changes felt worldwide. According to this, there is a need to deal with urban warming and its harmful effects, especially as more people move to cities.

Addressing Egypt's efforts in implementing the SDGs, most of the projects and initiatives currently being executed focus on the country's challenges, such as poverty, education, agriculture, energy, and urbanisation. Egypt should follow a comprehensive approach that combines the SDGs, SDSs, and Vision 2030, as this generates a better outcome from implementing their initiatives separately.

The Egyptian urban development process management is distributed among numerous government and local authorities administrative entities. These local authorities lack the needed resources and capacities, specifically in decision-making. All of the aspects mentioned before are expected in most Egyptian cities. Accordingly, implementing the SDGs can help in guiding the overall development process. The Egyptian government has committed to implementing the SDGs in the national Vision 2030 Agenda. However, local authorities are still far from comprehending the SDGs, including them in the development strategies.

Egypt successfully implemented the SDGs despite all the mentioned challenges, considering its current economic and regional unrest challenges. Moreover, achieving the targets of the SDGs and the national Vision 2030 should be supported by an integrated, long-term approach that would overcome the local authorities' shortfalls and ensure genuine community participation. Finally, how cities deal with the challenges of the new millennium and the problems we face today will depend on how they use their resources, how they prioritise them, and how they plan their cities for the next decade.

Proper urban planning can also play a vital role in mitigating the UHI effect. Yamamoto (2006) has described an urban planning approach situated on the bank of the river. His suggestion is to build the buildings to create a winding path for excellent airflow from the river into the city. If buildings are made parallel to the direction of the river, no airflow will occur into the city. If the buildings are positioned at a 450 angle, the wind will get a channel if it flows only in one direction. But it flows from the opposite direction, it cannot get away inside the city. However, airflow will occur if the buildings are perpendicular to the river. He also mentioned that this option is expected to play a major role in the fore coming days and it deserves further attention. In other types of cities, it is expected that if there is sufficient amount of free space and channel to circulate the wind, it will help to minimise the effect of the urban microclimate.

## REFERENCES


Adhikari, S. (2018, March 6). *Urban heat island, its effects and mitigation measures*. Retrieved from Public Health Notes: https://www.publichealthnotes.com/

Alchapar, N. L., & Cantón, N. A. (2016). Classification of building materials used in the urban envelopes according to their capacity for mitigation of the urban heat island in semiarid zones. *Energy and Building*, *69*, 22–32. doi:10.1016/j.enbuild.2013.10.012

Appleton, J. (2020, December 3). *How the urban heat island effect is harming our cities*. Retrieved from Bee Smart City: https://hub.beesmart.city/

Asi, Y. M. (2021, July 30). *Challenges facing sustainable development goals in Arab States*. Retrieved from Arab Center Washington DC: https://arabcenterdc.org/

CAPMAS. (2017). *Monitoring and follow up SDGs in Egypt*. CAPMAS.

David, K., & Bohl, T. L. (2018). Sustainable Development Goals Report: Egypt 2030. United Nations.

Devisscher, T., Konijnendijk, C., Nesbitt, L., Lenhart, J., Salbitano, F., Cheng, Z. C., & Bosch, M. v. (2019). SDG 11: Sustainable cities and communities – Impacts on forests and forest-based livelihoods. In P. Katila & C. J. Colfer (Eds.), *Sustainable development goals: Their impacts on forests and people* (pp. 349–385). Cambridge University Press. doi:10.1017/9781108765015.013

Egyptian Streets. (2020, October 26). *Egypt ranks first regionally for sustainable development goals assessment*. Retrieved from Egyptian Streets: https://egyptianstreets.com/

El-Magd, I., Ismail, A., & Zanaty, N. (2016). Spatial variability of urban heat islands in Cairo City, Egypt using time series of Landsat satellite images. *International Journal of Advanced Remote Sensing and GIS*, *5*(3), 1618–1638. doi:10.23953/cloud.ijarsg.48

Elmarakby, E., Khalifa, M., Elshater, A., & Afifi, S. (2022). Tailored methods for mapping urban heat islands in Greater Cairo Region. *Ain Shams Engineering Journal*, *13*(2), 101545. doi:10.1016/j.asej.2021.06.030

Elshater, A., Abusaada, H., Alfiky, A., El-Bardisy, N., Elmarakby, E., & Grant, S. (2022). Workers' satisfaction vis-à-vis environmental and socio-morphological aspects for sustainability and decent work. *Sustainability*, *14*(3), 1–25. doi:10.3390u14031699

Eurostat. (2020, May). *SDG 11 - Sustainable cities and communities*. Retrieved from Eurostat Statistics Explained: https://ec.europa.eu/

Farshid Aram, E. H. (2019). Urban green space cooling effect in cities. *Heliyon*. PMID:31008380

Fayed, L., Elshater, A., & Rashed, R. (2020). Aspects of regenerative cities. In S. Kamel, H. Sabry, G. F. Hassan, M. Refat, A. Elshater, A. S. Elrahman, . . . R. Rashed (Eds.), Architecture and Urbanism: A Smart Outlook. doi:10.1007/978-3-030-52584-2_22

Filho, W. L., Wolf, F., Castro-Díaz, R., Li, C., Ojeh, V. N., Gutiérrez, N., ... Böneck, J. (2021). Addressing the urban heat islands effect: A cross-country assessment of the role of green infrastructure. *Sustainability*.

Griloa, F., Pinhoa, P., Aleixoa, C., Catita, C., Silva, P., Lopes, N., ... Branquinho, C. (2020). Using green to cool the grey: Modelling the cooling effect of green spaces with a high spatial resolution. *The Science of the Total Environment*, *724*, 138182. Advance online publication. doi:10.1016/j.scitotenv.2020.138182 PMID:32408445

Hideki Takebayashi, M. M. (2020). Background and purpose. In *Adaptation Measures for Urban Heat Islands*. doi:10.1016/B978-0-12-817624-5.00001-4

Mirzaei, P. A., & Haghighat, F. (2010). Approaches to study Urban Heat Island – Abilities and limitations. *Building and Environment*, *45*(10), 2192–2201. doi:10.1016/j.buildenv.2010.04.001

Mohajerani, A., Bakaric, J., & Jeffrey-Bailey, T. (2017). The urban heat island effect, its causes, and mitigation, with reference to the thermal properties of asphalt concrete. *Journal of Environmental Management*, *197*, 522–538. doi:10.1016/j.jenvman.2017.03.095 PMID:28412623

Nuruzzaman, M. (2015). Urban heat island: causes, effects and mitigation measures - a review. *International Journal of Environmental Monitoring and Analysis*.

Ogujiubm, K. K., & Ehigiamusoe, K. U. (2013). The challenges and implications of sustainable development in Africa: Policy options for Nigeria. *Journal of Economic Cooperation and Development*.

SDG 11: Sustainable Cities and Communities. (2021). Retrieved from SDG 11: Sustainable Cities And Communities: https://in.one.un.org/page/about-us/

Seto, K. C., Güneralp, B., & Hutyra, L. R. (2012). Global forecasts of urban expansion to 2030 and direct impacts on biodiversity and carbon pools. *Proceedings of the National Academy of Sciences of the United States of America*, *109*(40), 16083–16088. doi:10.1073/pnas.1211658109 PMID:22988086

Shishegar, N. (2014). The impacts of green areas on mitigating urban heat island effect: A review. *The International Journal of Environmental Sustainability*, *9*(1), 119–130. doi:10.18848/2325-1077/CGP/v09i01/55081

Takase, C. (2018). *Implementing SDG 11 – key elements, challenges and opportunities*. The United Nations Centre for Regional Development.

Tan, N. H., Kolokotsa, D. D., & Takebayashi, H. (2021, January 26). *Greenery as a mitigation and adaptation strategy to urban heat*. Retrieved from Nature reviews earth & environment: https://www.nature.com/

The Global Goals. (2015). *The Global Goals*. Retrieved from The Global Goals for Sustainable Development: https://www.globalgoals.org/

The Ministry of International Cooperation. (2016). *Egypt national review report*. The Ministry of International Cooperation.

The Ministry of Planning, Monitoring, and Administrative Reform. (2018). Egypt's national voluntary report 2018. Cairo: The Ministry of Planning, Monitoring and Administrative Reform.

The Ministry of Planning and Economic Development. (2021). *Egypt'S 2021 voluntary national review*. The Ministry of Planning and Economic Development.

The Ministry Of Planning and Follow-up and Administrative Reform. (2016). *Sustainable development strategy (SDS): Egypt vision 2030*. The Ministry Of Planning and Follow-up and Administrative Reform.

UN Climate Change News. (2021, August 9). *Secretary-General's Statement on the IPCC Working Group 1 Report on the Physical Science Basis of the Sixth Assessment*. Retrieved from United Nations Climate Change: https://unfccc.int

United Nations. (2015). *Sustainable Development Goals*. Retrieved from United Nations: https://www.un.org/

United Nations. (2015). *Take Action for the Sustainable Development Goals*. Retrieved from Sustainable Development Goals: https://www.un.org/

United Nations. (2016). *Sustainable Development Goals*. Retrieved from Take Action for the Sustainable Development Goals: https://www.un.org/

United Nations. (2017). *Transforming our world: the 2030 Agenda for Sustainable Development*. United Nations.

United Nations. (2019). *The Sustainable Development Goals Report 2019*. United Nations.

United Nations Environment Programme. (2016). *Healthy environment, healthy people*. United Nations.

U.S. Environmental Protection Agency. (2016). *U.S. Environmental Protection Agency*. Retrieved from Reduce Urban Heat Island Effect: https://www.epa.gov/

World Meteorological Organization (WMO). (2021). *Urban heat island*. Retrieved from World Meteorological Organization (WMO): https://community.wmo.int/

# Chapter 3
# Dissecting Urban Heritage Morphology as a Response to Microclimate:
## Kampung Kauman, Indonesia

**Melvina Pramadya Puspahati**
*University of Stuttgart, Germany*

## ABSTRACT

*This chapter discusses the historic areas that are living habitats that grew before the science of architecture and urban design happened in a formal way. In the case of the Kampung Kauman neighbourhood area, the area is a result of an agglomeration of residents with similar job types as Batik makers. However, the district was designed by an indigenous Javanese planner or Kalang who has a responsibility to design and build the area. The method used a combination of computational climate simulation to investigate outdoor spaces, surface radiation. Further, this chapter conducted an ethnographic study qualitatively through field surveys. The results captured people's activities at different times during the day. The results also indicated that most of the Javanese buildings can consider climate as responding to buildings and public spaces. The concluding remarks identified the relation of the urban morphology from indigenous Javanese culture with people activities through spatial investigation.*

## INTRODUCTION

Cultural heritage is an impression of the civilisation developed by a community and passed on from generation to generation, including customs, practices, places,

DOI: 10.4018/978-1-6684-2462-9.ch003

objects, artistic expressions, and values. Cultural heritage is associated with expressing either intangible or tangible cultural heritage (International Council on Monuments and Sites (ICOMOS), 2011). Cultural heritage is an outcome of a tangible representation of the value system, beliefs, traditions, and lifestyle as part of human activity (Ashrafi, Kloos, & Neugebauer, 2021). The architectural heritage in some communities is exposed to environmental changes like climate changes and urban heat island due to human activities that might affect the value of these assets (Winter, 2016). This challenge is the focus of this research.

In literature, heritage is divided into concrete forms of origin such as artefacts, buildings, or landscape architecture (Fibiger, 2015). However, the tangible heritage is inextricably bound up with intangible heritage (Feather, 2006). Meanwhile, intangible heritage includes voices, values, tradition, oral history, cuisine, clothing, forms, skills, religious ceremonies, performing arts, and storytelling.

The history of batik is intertwined with cultural values, craftsmanship, and the development of agglomeration communities. Indonesia's batik was officially added to the UNESCO intangible cultural heritage of humanity in 2009 (Steelyana, 2012). This activity has been internationally recognised as part of human civilisations' historic fabric. In addition to intangible cultural heritage, the existence of the batik industry also has a significant influence on urban agglomeration through the "Kampong" community residential complex typology (Poon, 2020). "Kampung" is a type of settlement in Southeast Asian countries such as Malaysia, Indonesia, and Singapore. Each kampong has a traditional leader called a penghulu, or village chief.

Moreover, in architecture and master planning design, Javanese Kampung has a Kalang who is responsible for building and neighbourhood planning. In Javanese culture, it has five types of Kalang: Kalang Kaba as an expert of planning and design, Kalang Bladong as an expert for wood quality selection, Kalang Obong as a land clearing expert, and Kalang Adeg as construction expert and Kalang Breg as building demolished expert. Hence, understanding the principle of Javanese architecture would be a starting point to preserve a living heritage and the relation to climate adaptation.

The research case is the transformation of Kampung Kauman as a living heritage. Surakarta is in Central Java. The Kampung has mentioned Indonesia's national origin based on Mayor Decree No.646/116/1/1997 on the building heritage sites' governance laws stipulation. The district has a mosque as its centre of spatial orientation. Kauman has the meaning of Quran student and is marked by old traditional Javanese and Indo-European architectural references, or detached. The area was provided for Quran students, but as modernity changed the main activity, it became a home industry complex. The batik industry is one of the economic activities to finance the Quran student complex. The batik industry has been going strong in Kauman for a long time, even during a terrible time in 1939. Then, in 1979, the business went bankrupt and has only recently been able to get back on its feet (Musyawaroh, Pitana, Masykuri, & Nandariyah, 2018).

Kauman is a district typology that grows organically because of thematic job activity as a batik crafter. Some of the ancient houses in Kauman have been transformed into commercial buildings such as shops, convenience stores, bookstores, florists, jewellery stores, restaurants, and office stores. Kauman is a community settlement with 3.553 inhabitants in 2016 (Musyawaroh, Pitana, Masykuri, & Nandariyah, 2018). Nowadays, most residents in the district are crafters and traders (719 people), construction workers (111 people), industry labourers (155 people), entrepreneurs (348 people), transportation (65), pensioners (74 people), civil servants (49 people) and others (1,341 people). In 2006, the district had only eight batik industries, but now it is up to 100 entrepreneurs in the home industry. During the transformation into the modern era, the community inside Socionic has been actively involved in revitalising and reviving their area by establishing the Associate Socionicurism Kampong Batik Kauman. This association collaborates with the academic institutions from the Architecture Program at Sebelas Maret and private institutions (Association of Experts Management and Association of Transportation Entrepreneurs).

The results from Musyawaroh et al. (2018) are considered sources of data related to the figures and ground of Kauman and green open space in Kauman. Much literature has supported this research on microclimates and people's behaviour to respond to thermal conditions (Winter, 2016). Moreover, research on the sustainable revitalisation of Kampong Kauman with reflection on actual data through statistical analysis of the cultural heritage (Musyawaroh, Pitana, Masykuri, & Nandariyah, 2018). However, there are limited studies on the historical aspects of Kampung Kauman, Surakarta as a living heritage in Javanese culture backgrounds, such as journals related to Javanese planners' responsibility and principle.

This chapter tackles the challenge of the modernity movements in today's cities that affect the historical monuments and heritage areas, which could lead to the possible degradation of heritage assets' originality. Meanwhile, modernity also impacts the transformation of building functions inside the heritage area, such as hotels and commercials inside the heritage area. Kampung Kauman is a heritage area where residents still preserve activities and keep the buildings in their original condition.

This research attempts to answer the following central equations: How are the social activities in outdoor space inside the Kampung Kauman heritage area reflected in microclimate conditions? What is the street canyon component that impacts social activities in outdoor space?

The aim of the present research is threefold. The first is to identify the current activity in outdoor space inside Kampung Kauman throughout the day, focusing on the change in activities responding to different times of thermal and air temperature conditions. The second aim is to record people's responses to climate phenomena, such as clothing and building colour. Finally, the third aim is to identify the component of the street canyon that influence microclimate, such as the relation of street canyon width, vegetation size, etc.

Kampung Kauman is in a tropical climate zone with an annual average humidity of around 74% (World Weather & Climate Report, 2021). Kampung Kauman urban fabric has the characteristic of narrow street canyon, and people mostly use the street as a place for socialisation rather than a big plaza. Hence, a hypothesis is raised that the microclimate condition in Kampung Kauman influences the resident activities in the outdoor spaces.

This research is divided into three main sections. The first section figured out the case study analysis to identify the physical attributes of outdoor space and activities in outdoor space throughout the day. The physical component of outdoor space includes the width of the street canyon, earth surface material, and vegetation. Second, the literature review attempted to present the concept of microclimate, tangible components that influence microclimate and standard of thermal comfort for tropical climate location. Finally, the last section provides lessons learned from the case studies.

## LITERATURE REVIEW

### Urban Morphology

Morphology is a word that is firstly described in biology terms as representative of repeating forms of organisms, and urban is a word to describe a place of living habitat related to the number of population and an area with typical jobs in non-agricultural sectors such as service, industrial (Kropf, 2017). Urban morphology is a comprehensive study of human settlements that includes the structure and the process of their arrangement and transformation (Bobkova, Berghauser, & Marcus, 2021; Elshater, The Predicament of Post-Displacement Amidst Historical sites: a design-based correlation between people and place, 2019; Kropf, Plots, property and behaviour, 2018). Urban morphology explains a solid descriptive language of the built environment. It helps to understand the characteristics of the city and the diversity of human culture as expressed in local forms of various human settings, both in the modern era and historically. In her book, Karl (2017) mentioned that "in a combination of a range of methods and techniques, urban morphology supports tools for investigating and assessing urban environmental performance as a habitat and the suitability for different purpose of development". Moreover, Urban morphology is a potential descriptive explanation for generating other studies such as energy use, microclimate, acoustic, movement, land use and land value (Kropf, 2018).

Urban morphology refers to the physical elements of settlements or habitants' areas that include the structure and shape of the city, including streets, squares, block massing, plots, and buildings (Oliveira, 2020). However, the study by Jenks and Dempsey (2005) mentioned that urban morphology is related to non-physical aspects such as density, which is strongly linked with the configuration of social environment and interaction within neighbourhoods. Hence, Urban morphology generally comprises tangible and intangible characteristics, including size, shape, scale, density, land use configuration, building type, urban block, and distribution of urban space.

Urban morphology classification comes here based on shaping a metropolitan development area. Some scientists have classified urban morphology into planned and unplanned development. The planned development comes from how an urban planner or designer sets up the urban form. They help to design the settlement areas. On the other hand, spontaneous development usually has a solid connection to how people in a settlement use their knowledge and limited resources to make or arrange an urban morphology (Pacione, 2009). So, it does not mean all unplanned settlement classifications are built without a plan. People who live in an area where there isn't a lot of planning could be the people who make decisions about how to deal with the natural environment with limited skills and resources. Consequently, the terminology of planned and unplanned urban morphology is biased. In addition, urban morphology has also been associated with urban patterns as a descriptive language to classify typologies of urban morphology based on physical aspects divided into building blocks, streets, and open space areas.

The urban forms have been divided into geometric and organic shapes. Geometric forms are associated with linear, grid, radial, and combining that configuration into the development area. At the same time, organic forms are mainly the result of tackling natural topographic conditions or natural elements such as rivers, mountains etc. However, the organic form of urban morphology is possibly associated with the sprawl of development in unplanned development cases.

## Micro-Climate and Urban Heat Islands

Microclimate terminology was established by German and British meteorologists and geographers in the first half of the 20$^{th}$ century. It describes that micro-climate is the climate in the air layer as about two metres above the ground (Elmarakby, Khalifa, Elshater, & Afifi, 2022). The idea of microclimate is not only about implementing a standard but rather for creating thermal comfort depending on the on-site situation and people approached. Hence, microclimate is a site-specific thermal zone influenced by physical and thermodynamic characteristics (Roesler & Kobi, 2018). Yang, Lin, and Li (2018) say it changes because of things like temperature and moisture and the amount of light it gets and how cloudy or clear it is.

Three fundamental elements of microclimatology for architecture and landscape are the microclimate of the plant cover, the topography, and the buildings. The microclimate is generated based on the urban morphology of the area. It is mentioned that microclimate can impact to 17% of an increase in cooling degree-day and 7% of a decrease in heating degree-day on average (Hosseini, Javanroodi, & Nik, 022). Microclimate simulation produces meso-climate data parameters with physical condition analysis variables. The energy performance and bioclimatic are consequent by microclimate conditions (Elshater, et al., 2022). Morphological characteristics including urban form, Albedo variable, vegetation density, solar gain, solar reflectivity, and wind speed are those parameters to measure microclimate conditions.

The modelling tools have a role in supporting evaluating building energy performance or thermal condition. Microclimate modelling is data set for assessing microclimate conditions on site-specific locations. The volumetric data consists of the land surface, building massing, external elements of the case study area such as a tree, façade material etc.

The baseline climate data set is used as a parameter in microclimate analysis. The weather data consisted of dry-bulb temperature, dew point temperature, relative humidity, atmosphere pressure, direct solar radiation, horizontal solar radiation, global illuminance, wind direction, perception dept and Albedo surface. The data mostly takes a sample from airport location or city centre point for significant city cases of places.

# METHOD AND MATERIAL

## Research Design

The research method used a combined approach to record quantitative data. This methodology for measuring microclimate in Kampung Kauman is conducted by parametric environmental-climate analysis through Rhino-Grasshopper Ladybug 1.2 tools. Besides, the primary data collection is undertaken by assistant research and a desktop research study for secondary data—literature review and journal study—to get the physical context of the case study.

The parametric environmental-climate simulation was calculated the outdoor thermal climate index. It is an interactional standard for outdoor temperature feels, which result from humidity data, air temperature data, mean radiant temperature data, and wind velocity data. In addition, mean radiant temperature data for outdoor shortwave solar shining directly onto people. The component applies the SolarCal model of ASHRAE-55 to estimate the effects of shortwave solar and a simple sky exposure method to determine longwave radiant exchange. The specific date will be selected as a critical analysis time, such as the highest air temperature or the highest humidity index.

This research would be beneficial from a climate perspective to propose a new typology of residential development in the modern era with comforting micro-climate aspects.

## Data Collection

### Case Study Description

Kampung Kauman is in central Java, Indonesia. The Kampung is part of the Surakarta city administrative area and the Kasusnan Surakarta Palace Complex. Kampung Kauman is a living heritage which is a settlement complex for Abdi Dalem Ulama Kreraton or palace leader servant, Batik businessman and a place for local ceremonial such as *Sekaten* and *Grebeg* ceremony (Musyawaroh, Pitana, Masykuri, & Nandariyah, 2018). The Kampung was mainly designated for religious student "*santri*" and then transformed into a village for a local fabric (*Batik* fabric[1]) industrial complex. The settlement configuration in terms of architectural and urban design approach has a strong influence from the Islamic mezzo space pattern (Nurjayanti, 2019) and Javanese architecture principles. The mosque has a role as a development orientation surrounded by settlement functions such as houses, Islamic school/ madrasah, community building "*langar*", and business. While the influence of Javanese architecture philosophy appears on the local architect knowledge, "*Kalang*" organise the traditional building and settlement based on cosmological layout. Understanding Javanese cosmological layout is a concept based on an assumption of the "other," that is orderly and mystical who live in a harmonic realm (Ofita, 2018). The spatial pattern of Kampung Kauman, Surakarta is orderly oriented to the north-east, hence the opening of buildings such as doors and windows toward the north or east, for the purpose of airflow (Nurjayanti, 2019). The Kampung Kauman settlement has a grid spatial housing pattern and a narrow corridor.

The site area of Kampung Kauman is 20.497 hectares, primarily buildings and pavement covering the land surface. The building is mainly a low-rise building, which means it has under four floors. Figure-ground map (Figure 1) describes that around 40% of the land surface of the Kampung Kauman is green open space, as amount 438.249 m$^2$, including private and public green open space. At the same time, 62,75% of the building construction area is 128.622 m$^2$.

*Figure 1. Kampung Kauman area, Source: Musyawaroh 2016*

## Geographical Context and Macro Climate Characteristic

The Kampung Kauman is positioned on 7°34’22.3”S 110°49’31.8”E, which is in the equatorial line, mainly in the tropical climate zone. It has an equal season distribution for the year. The duration of the sun hour per day is almost identical every day as it is around a 12-hour sun hour period (Figure 2). Based on the Köppen climate classification, Kampung Kauman is in the monsoon climate zone. The climate has been affected by the high pressure of monsoon winds from the south during the wet monsoon between September and April and by the high pressure of monsoon winds from the north during the dry monsoon between May and August. The dry monsoon season is caused by the Intertropical Convergence Zone (ITCZ) wind from the south. The specific characters are high precipitation as around 1000mm per year during the wet season, resulting in high humidity above 60% average per month.

## Microclimate Data – Central Java

The sample climate data was taken from a sample in Madiun, 48 miles east of Surakarta. The reason for using climate data from Madiun is that there is no climate

*Figure 2. Indonesia climate classification based on Köppen–Geiger Climate Classification*
*Source: Adapted from Kottek, 2006*

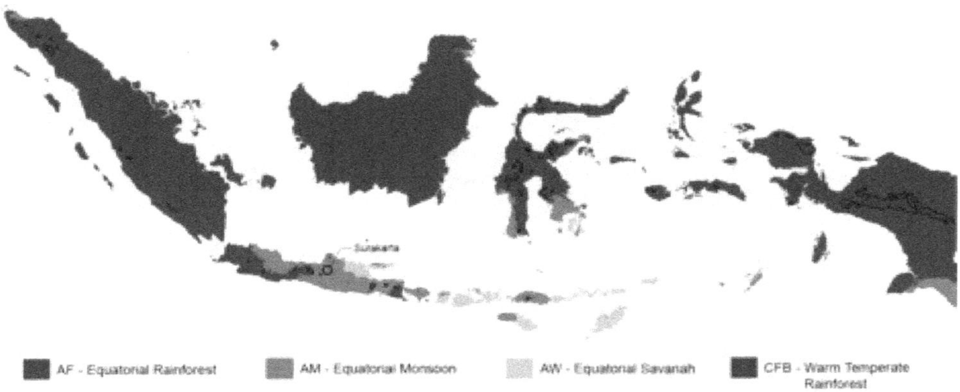


data available for Surakarta. However, both Madiun and Surakarta are in a tropical monsoon climate zone.

Based on climate data, the sun path position during the daytime is between 60°–90°. Especially at noon, when in the high radiation, the sun position is perpendicular 90°. The highest wind speed is coming from the south, as amount 12 m/s. The average wind speed is around 6 m/s per day. The temperature range is between 26° - 31° Celsius, which is comfortable throughout the day. However, from January to May and November to December, there is a lot of humidity, up to 80%, making people sweat.

Overall, the leading climate challenge for central Java is the perpendicular of direct sun radiation, high humidity, and high precipitation. The critical time for the most increased solar radiation is on the 21st of June when the sun is positioned at 90°. At the same time, the highest humidity is on the 1st of May, when the humidity reaches 100%. Meanwhile, the daily average of precipitation is around 80mm; it can be categorised as high precipitation.

## Sampling Case Study Area

A site survey was conducted to investigate the building configuration and additional outdoor elements such as tree and street canyon dimensions. Figure 5 shows the survey images.

Based on the survey, the building block configuration in Kampung Kauman is an organic configuration divided by a street grid. There is no specific pattern of building configuration. However, the street canyon dimension has a little distance

*Figure 3. Central Java climate data sample Madiun*
*Source: https://www.ladybug.tools/epwmap/*

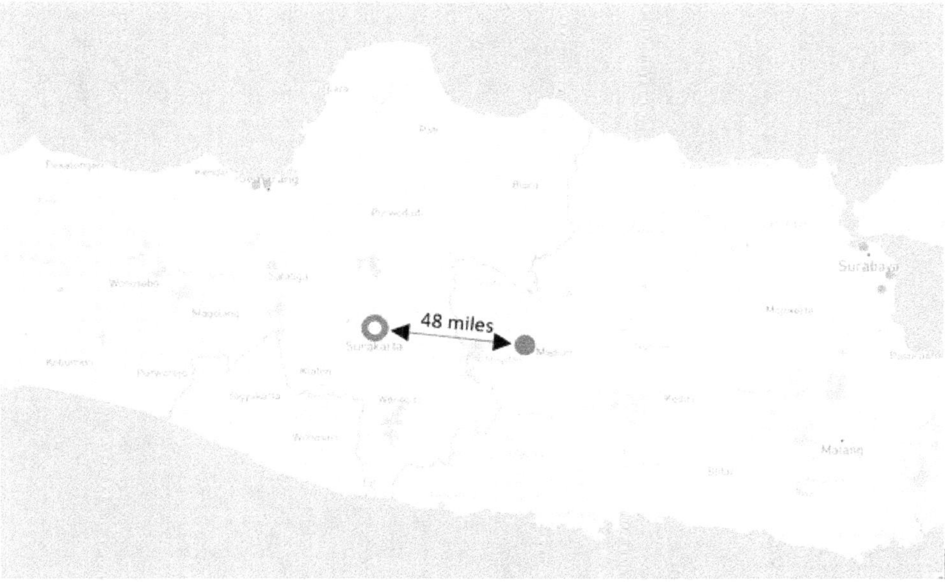


of about 1 meter to 5 meters. The buildings high in the sample case study are in the range 1 floor (4,5 meters include roof high), 2 floors (6 meters include roof high) and three floors (10 meters include ceiling high). The survey was taken on 2 January 2022 around 12 pm – 3 p during the daytime.

## RESULTS, DISCUSSION AND CONCLUSION

The simulation is taken at a critical time considering specific climate challenges in Central Java. The first climate challenge at the highest humidity, up to 90%, is May. It has resulted in the thermal feels range at the highest humidity at about 31° – 36° Celsius. The second selected period of the analysis is on the most increased solar radiation, which the thermal feels is reached 27° up to 32° Celsius. The last selected date is taken in the highest air temperature condition. The thermal feels are about 33° – 40° Celsius.

The area with high surface exposure to direct radiation increases thermal feel. Figure 6 shows that the site has a 3° Celsius temperature than the narrow surface. However, the building heights and street distance ratio influence thermal feels calculation. Compared to air temperature climate data, the building configuration

*Figure 4. Central Java climate data sample location Madiun*
*Source: https://climate.onebuilding.org/WMO_Region_5_Southwest_Pacific/IDN_Indonesia/JW_Jawa/IDN_JW_Madium-Iswahyudi.AP.968810_TMYx.zip*

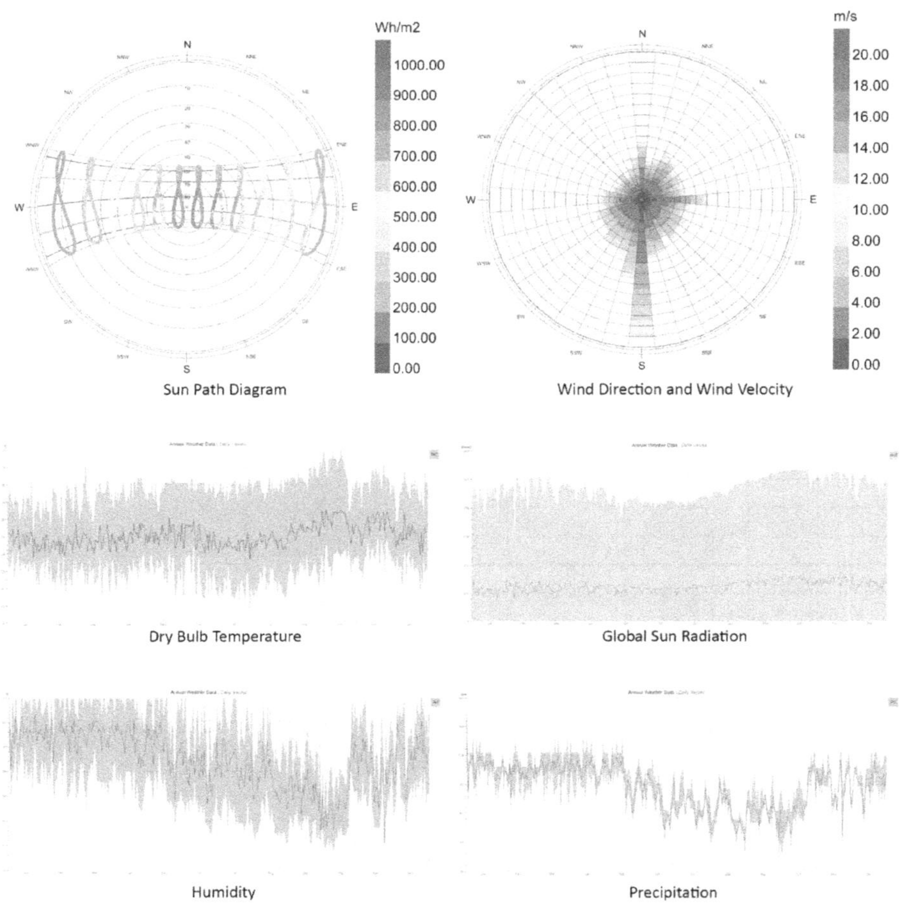


in Kampung Kauman is also increasing thermal feels as around 2° Celsius higher. The selection of street material would be another factor that possibly influenced the result.

Based on the calculation and simulation of outdoor microclimate in the sampling case study area, it can be concluded that a narrow street canyon in a tropical climate context can decrease thermal feels. The analysis results influence the climate parameters, air temperature, relative humidity, mean air temperature, and wind velocity. This study mentioned that Kampung Kauman had been affected by the Javanese principle, mainly the building oriented to north-south. The urban

*Figure 5. Site survey image for the compilation of the sample of the case study analysis, source: The author's*

morphology of building massing configuration in Kampung Kauman cannot be identified considering climate context because no specific pattern is found.

In addition, a narrow street canyon entirely in the shadow is suitable for an outdoor activity to support zero-carbon mobility, such as walking and cycling. Considering Musyawaroh's research about sustainable revitalisation in the cultural heritage of Kampong Kauman and Surakarta supported by spatial analysis, revisioning the regeneration of Kampung Kauman should be viewed as the option of no carbon mobility service as responding the street.

*Figure 6. Simulation result of universal thermal index the calculation for outdoor in critical climate time, source: The author's*

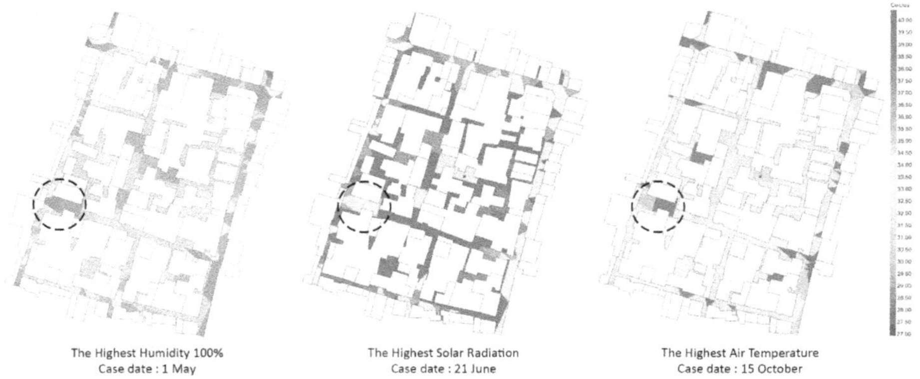


## ACKNOWLEDGMENT

The authors would like to thank (1) Professor Abeer Elshater and Hisham Abusdaada as editor team, (2) Musyawaroh Sugito as a researcher of Kampung Kauman Surakarta, (3) Anisa H Alamanda as a research assistant, (4) DAAD scholarship.

## REFERENCES

Ashrafi, B., Kloos, M., & Neugebauer, C. (2021). Heritage impact assessment, beyond an assessment tool: A comparative analysis of urban development impact on visual integrity in four UNESCO World Heritage properties. *Journal of Cultural Heritage*, *47*, 199–207. doi:10.1016/j.culher.2020.08.002

Bobkova, E., Berghauser, P. M., & Marcus, L. (2021). Towards analytical typologies of plot systems: Quantitative profile of five European cities. *Environment and Planning. B, Urban Analytics and City Science*, *48*(4), 604–620. doi:10.1177/2399808319880902

Elmarakby, E., Khalifa, M., Elshater, A., & Afifi, S. (2022). Tailored methods for mapping urban heat islands in Greater Cairo Region. *Ain Shams Engineering Journal*, *13*(2), 101545. doi:10.1016/j.asej.2021.06.030

Elshater, A. (2019). The Predicament of Post-Displacement Amidst Historical sites: A design-based correlation between people and place. *Heritage & Society*, *12*(2-3), 85–115. doi:10.1080/2159032X.2021.1879355

Elshater, A., Abusaada, H., Alfiky, A., El-Bardisy, N., Elmarakby, E., & Grant, S. (2022). Workers' Satisfaction vis-à-vis Environmental and Socio-Morphological Aspects for Sustainability and Decent Work. *Sustainability*, *14*(3), 1–25. doi:10.3390u14031699

Feather, J. (2006). Managing the documentary heritage: issues for the present and future. In G. Gorman (Ed.), *Preservation management for libraries, archives and museums* (pp. 1–18). doi:10.29085/9781856049214.002

Fibiger, T. (2015). Heritage erasure and heritage transformation: How heritage is created by destruction in Bahrain. *International Journal of Heritage Studies*, *21*(4), 390–404. doi:10.1080/13527258.2014.930064

Hosseini, M., Javanroodi, K., & Nik, V. M. (2022). High-resolution impact assessment of climate change on building energy performance considering extreme weather events and microclimate – Investigating variations in indoor thermal comfort and degree-days. *Sustainable Cities and Society, 78*, 103634. . doi:10.1016/j.scs.2021.103634

International Council on Monuments and Sites (ICOMOS). (2011). *ICOMOS World Report 2002-2003 on monuments and sites in danger.* Retrieved February 2022, from https://www.icomos.org/en/what-we-do/risk-preparedness/heritage-at-risk-reports/116-english-categories/resources/publications/210-icomos-world-report-2002-2003-on-monuments-and-sites-in-danger

Jenks, M., & Dempsey, N. (2005). *Future Forms and Design For Sustainable Cities.* Routledge.

Kottek, M., Grieser, J., Beck, C., Rudolf, B., & Rubel, F. (2006). World Map of the Köppen-Geiger climate classification updated. *Meteorologische Zeitschrift (Berlin)*, *15*(3), 259–263. doi:10.1127/0941-2948/2006/0130

Kropf, K. (2017). *The handbook of urban morphology.* John Wiley & Sons Ltd. doi:10.1002/9781118747711

Kropf, K. (2018). Plots, property and behaviour. *Urban Morphology: International Seminar on Urban Form, 22*(1).

Musyawaroh, M., Pitana, T. S., & Masykuri, M., & Nandariyah. (2018). Sustainable revitalisation in Cultural Heritage Kampong Kauman surakarta supported by spatial analysis. *IOP Conference Series. Earth and Environmental Science*, *123*(1).

National Centers for Environmental Information. (2021). *Global Climate Report - Annual 2021*. NOAA. Available online at https://www.ncdc.noaa.gov/sotc/global/202113

Nurjayanti, W. (2019). Micro and mezzo space pattern in Kampung Kauman Solo. AIP Conference Proceedings, 2114(1), 40015. doi:10.1063/1.5112444

Oliveira, V. (2020). Urban Morphology. In R. Dilworth (Ed.), *Urban Studies*. Oxford University Press. doi:10.1093/obo/9780190922481-0004

Pacione, M. (2009). *Urban Geography A Global Perspective*. Routledge. doi:10.4324/9780203881927

Pitana. (2018). *The Architecture of Omah as an Expression of the Beauty of Javanese Culture*. Academic Press.

Poon, S. (2020). Symbolic resistance: Tradition in batik transitions sustain beauty, cultural heritage and status in the era of modernity. *World Journal of Social Science*, *7*(2), 1–10. doi:10.5430/wjss.v7n2p1

Purwani, O. (2018). Cosmological Layout and Hierarchy. *Colonial Narrative and Structure in Javanese Palaces, 294*. Available online at https://www.academia.edu/38186759/COSMOLOGICAL_LAYOUT_AND_HIERARCHY_Final_pdf

Roesler, S., & Kobi, M. (2018). *The urban microclimate as artifact: Towards an architectural theory of thermal diversity*. Birkhauser. doi:10.1515/9783035615159

Steelyana, E. (2012). Batik, A Beautiful Cultural Heritage that Preserve Culture and Supporteconomic Development in Indonesia Authors. *BINUS Business Review*, *3*(1), 116. Advance online publication. doi:10.21512/bbr.v3i1.1288

Winter, T. (2016). Climate change and our heritage of low carbon comfort. *International Journal of Heritage Studies*, *22*(5), 382–394. doi:10.1080/13527258.2016.1157092

Yang, W., Lin, Y., & Li, C.-Q. (2018). Effects of landscape design on urban microclimate and thermal comfort in tropical climate. *Advances in Meteorology*, *2018*, 1–13. doi:10.1155/2018/2809649

## ENDNOTE

[1] A method (originally used in Java) of producing coloured designs on textiles by dyeing them, having first applied wax to the parts to be left undyed.

# Chapter 4
# Regenerative Approach for Solid Waste Management in Historic Centres

**Diana Barrera**
*University of Stuttgart, Germany*

**Yara Elmaghrabi**
*University of Stuttgart, Germany*

## ABSTRACT

*Solid waste management is considered one of the greatest challenges faced by historic centres due to the nature of their urban fabric and the changing frequency of users of the area, resulting in variability in the amount and composition of solid waste produced. Although historic centres are places where the Genius Loci of cities are appreciated, the implications of poor solid waste management led to the deterioration of these significant places. The application of holistic strategies contributes to keeping the urban identity alive while considering current social and economic aspects. This chapter focuses on exploring the potential of using the urban regeneration framework for solid waste management in historic centres. The discourse begins by defining the conceptual framework, followed by reflections on the historic centres of Naples and Cairo to explore how regenerative approaches can be used. Finally, the findings suggest a possible regenerative framework that could be used by different groups to solve problems with waste management in historic cities.*

DOI: 10.4018/978-1-6684-2462-9.ch004

## INTRODUCTION

The international agendas of the 21st century are often defined by climate change and environmental sustainability, especially in cities (Carter *et al.*, 2015). This is particularly relevant today, as world leaders and local communities call for a solution to the so-called "*waste culture*" (Connett., 2013). However, solid waste is a broader problem that affects human health and livelihoods, the environment and prosperity, and individuals and households (Kaza *et al.*, 2018). In the case of historical centres, despite their historical importance and the relevant role they play in the urban fabric of cities, they face several challenges that need to be addressed to offer a sustainable future to both the residents and visitors (Khan, 2015). One of the most visible challenges is the poor management of solid waste and how this situation has affected the tangible and intangible heritage of the historic centre and, ultimately, the whole city (Ciudin *et al.*, 2014).

The role of Solid Waste Management (SWM) is increasingly crucial for both daily life and the long and medium-term sustainability of historic centres. The challenge of efficient waste management in these centres demands locals, public authorities, and the SWM industry (Laso *et al.*, 2019). If this issue is not addressed holistically, environmental problems will increase, resulting in poor hygiene and safety conditions for the users of this area. Nevertheless, historical centres are given little consideration to the frameworks that are being introduced to provide a holistic solution to solid waste management (Gaber, 2019).

In recent years, scholars and academia have been moving away from the notion of sustainability towards a more comprehensive approach named *Urban Regeneration* to solve the current global challenges, specifically the degradation of the ecosystem. This approach throws the light that cities cannot be just sustainable anymore but also need to be regenerative to ensure prosperity while maintaining balance with the ecosystem (Fayed *et al.*, 2020). Accordingly, policies have been introduced to confirm the application of regenerative strategies in cities in aspects such as energy sufficiency, nature reintegration, water security, and waste-free culture (Xin (Cissy) Ma, 2015; Darinka, 2015; Girardet, 2014).

The question that has driven this research is: How can solid waste management in historic centres shift towards a regenerative approach? Two sub-questions complement this question to understand the status quo of SWM in historical centres: How does the traditional SWM system affect the tangible and intangible heritage? What are the gaps in the current SWM systems?

This chapter aims to answer these questions by examining approaches to SWM in historical centres and investigating the potential of regenerative frameworks to improve the status quo. The research is conducted by critiquing current approaches from the point of view of regenerative frameworks in the context of two case studies that were chosen for their great historical and cultural relevance both for the cities and for the Mediterranean region of which they are part: historic Cairo in Egypt and historic Naples in southern Italy.

The primary motivation for addressing this issue is its relevance in the global context and how a regenerative solution that improves conditions for locals and visitors can be achieved through an approach that considers both social and environmental elements toward a sustainable heritage in historic areas. To further investigate SWM in historical centres, desktop research is conducted to address the research questions stated above in the selected case studies.

The regenerative approach developed as a result of this chapter can be used by locals, public authorities, and the waste management industry to create strategies and action plans to address the problems related to SWM. Moreover, it would serve as a starting point for further research on how efficient SWM plans can contribute to the sustainability of the heritage in historic centres, prevent their physical and social deterioration, and enhance their development.

## DISCOURSE: THEORETICAL FRAMEWORK FOR SOLID WASTE MANAGEMENT

Solid Waste Management (SWM) can be defined as the collection, treatment and disposal of solid materials that have served their purpose and are no longer needed (RCRA, 1976). In the 21st century, the rapid transformation of cities has led to an increase in the amount and complexity of solid waste composition (Laso, *et al.*, 2019) (Figure 1). Both the rise in global population density and the consumption trends are resulting in higher production of solid waste, confronting the cities with the challenge of implementing a practical SWM framework (Danish, *et al.*, 2019). Consequently, it will be necessary for cities to link conceptual and technological advances to move towards the concept of regenerative cities.

Historic centres around the world are characterised by narrow, winding streets where, in most cases, people and vehicles cannot coexist together, causing traffic jams and altering the daily routine of citizens (Khan, 2015). To address this problem, many cities have pedestrianised and widened streets, making waste collection traditional methods, such as street bins and garbage trucks, more feasible. Moreover, in recent years, these areas have become commercial and leisure centres, combining banks, company headquarters and public institutions with shops, hotels and restaurants, generating large quantities of solid waste, which represents a challenge for cities to manage appropriately, and increases the pressure on the entire SWM system of the cities (Gidarakos, Havas, and Ntzamilis, 2006).

*Figure 1. Variation in the municipal solid waste (MSW) generation and composition due to the changes in the population. (Laso et al., 2019)*

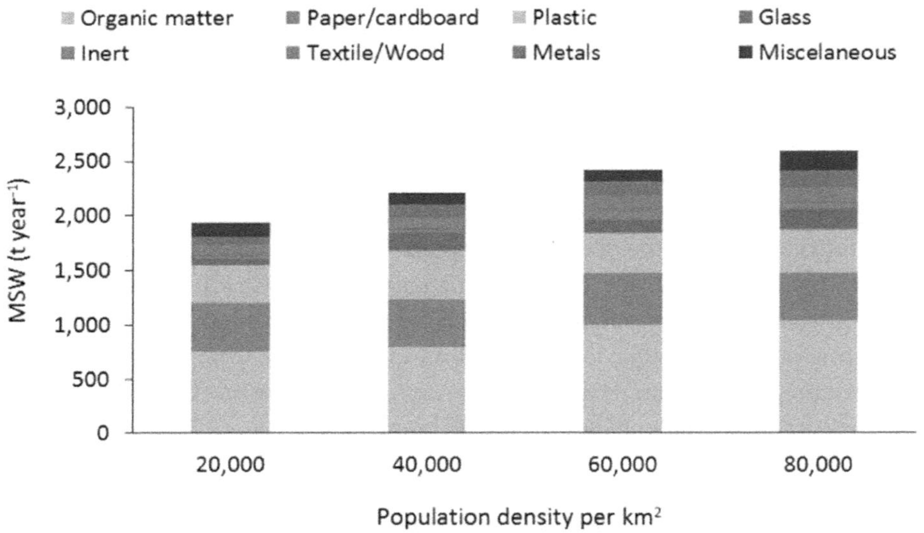


One of the characteristics of historic centres is that they are heavily frequented by tourists, a frequency which varies according to the season, and which is directly related to the variability of the composition of the solid waste generated. It is considered that the increase in population density, even if temporary, affects the composition of waste in these areas. In this sense, the need to find a suitable SWM system for historic areas is an issue that has opened the debate around the election parameters for the best collection system, as conventional street collection systems are not viable due to traffic and access restrictions. The most commonly used current system is the door-to-door collection, which generates significant direct greenhouse gas emissions from trucks (Punkkinen, *et al.*, 2012). Therefore, there is an urgent call for cities to find a greener option for SWM in historic areas under a circular economy approach, as waste could represent several opportunities as a source of income and energy source for local development (Gidarakos, Havas, and Ntzamilis, 2006). The inclusion of local actors in such a strategy will promote ownership of the place, as well as social, economic, and environmental sustainability.

The society also seems to be aware of the importance of separation and efficient collection of different types of waste. However, the current debate focuses on the most environmentally appropriate system without considering social aspects that could also be an advantage. Historic centres can be ideal for implementing strategies where monitoring and feedback from the local population is a key part of the proper management of these areas. Community-based waste management and circular economy are examples in this regard, as they are systems that are complemented by social and economic aspects, the inclusion of which can influence the balance towards one or the other management option (Laso *et al.*, 2019). In addition, an assessment of the social, economic, and environmental sustainability of collection systems could confirm the long-term benefits of the selected method.

## URBAN REGENERATION IN HISTORIC CENTRES

Urban change and transformation are inevitable due to rapid urbanisation and the constant introduction of new circumstances. This increases the pressure on the ecosystems and the available resources as human needs simultaneously grow. While some cities can adapt to those transformations, other cities have difficulties adapting, leading to urban decline. The urban decline has been identified as the reduction in employment opportunities and the loss of population. In the last decades, more and more cities have been subject to urban decline, which led to abandoned buildings, unemployment, and degradation in social life. The policies and strategies that deal with urban decline and urban transformations are grouped under "Urban Regeneration" (Tanrıkul & Hoşkara, 2019; Lang, 2005)

The UN-Habitat defined urban regeneration as an approach that "*brings back underutilised assets and redistributes opportunities, increasing urban prosperity and quality of life*" (UN-Habitat, n.d). In 2000, Roberts interpreted urban regeneration as a "*comprehensive and integrated vision and action which leads to the resolution of urban problems, and which seeks to bring about a lasting improvement in the economic, physical, social and environmental condition of an area that has been subject to change*" (Roberts, 2000: 17). On the other hand, the Priority Actions Program from the United Nations Environment Programme states that there are two ways to define urban regeneration based on the country's level of development. The first one is for the developed economies where "*the goal is to promote a "return to the city", revitalise the city centre, restore activity in a fiercely competitive international context, and implement initiatives to improve the quality of the environment operating in a wide sense towards a smart growth.*", the second one stated that in emerging economies "*qualitative initiatives have to be kept in mind, but must imperatively be aligned with quantitative requirements as well.*"(UNEP, 2004: 7).

A recent study states that urban regeneration is one of the tools used to achieve sustainability (Tanrıkul & Hoşkara, 2019). Accordingly, urban regeneration aims to account for the complexity of the urban dynamics, and instead of just focusing on the renovation and rehabilitation of the physical structures, urban regeneration targets restructuring the urban fabric, reintroducing the imageability of the city, and renewing the urban economy, while pursuing social interaction, local participation, and balance with the ecosystem (UNEP, 2004).

According to the Priority Actions Program, urban regeneration has several principles: location-specific as it is tailored specifically for the difficulties of a particular context while taking into consideration the global agenda; multi-temporal as it addresses the present needs while accounting for the future changes; and multi-dimensional as multiple stakeholders are involved in the process. Moreover, urban regeneration has four primary purposes in four different fields: the economic aspect, which includes reviving the economy and creating new job opportunities; the social element which aims to increase the urban housing and the local infrastructure; the environmental aspect, where the living conditions are developed while fighting pollution; and finally the cultural element that focuses on enhancing the urban tourism, enriching the physical heritage and the historic core of the city (UNEP, 2004).

Cities are dynamic structures with residential, social, civil and leisure amenities, and it is no surprise that the city centres play an essential role in how successful a city function. When that centre is historic, it adds another tangible and intangible heritage layer. It holds within it the city's identity, memory, and soul, all while trying to keep up with the technological advances and the constant changes in social and economic needs. However, the preservation of historical centres and their urban tissue have been one of the many challenges that cities face as they grow and transform. Additionally, ensuring the sustainable conservation of historic centres would keep the urban identity alive and improve the city's economic and social aspects (Tanrıkul & Hoşkara, 2019).

As a result, in 2019, Tanrıkul & Hoşkara (2019) developed an urban regeneration framework for historic centres in the Mediterranean Region. The authors stated that this framework will help apply urban regeneration processes in historic centres in general but for Mediterranean cities specifically. The framework (Figure 2) can be used during the design, implementation and evaluation for projects that seek urban regeneration.

According to recent studies, policies that encompass urban regeneration strategies need to be integrated into the national urban policies. Furthermore, aspects including energy sufficiency, nature reintegration, water security and waste recycling can be adapted. Solid waste management is one of the significant problems that historic centres face and applying a "waste-free" vision where policies such as developing enterprises that can process organic waste into soil enhancing materials can be implemented (L. Fayed et al.,2020; (Xin (Cissy) Ma, 2015; Darinka, 2015; Girardet, 2014). Consequently, shifting from a linear economy to a circular economy is necessary to move towards a more restorative and regenerative design. This paradigm shift will require the involvement of multiple stakeholders and different sectors within the city (MacArthur, n.d).

*Figure 2. Urban regeneration framework (Tanrıkul & Hoşkara, 2019)*

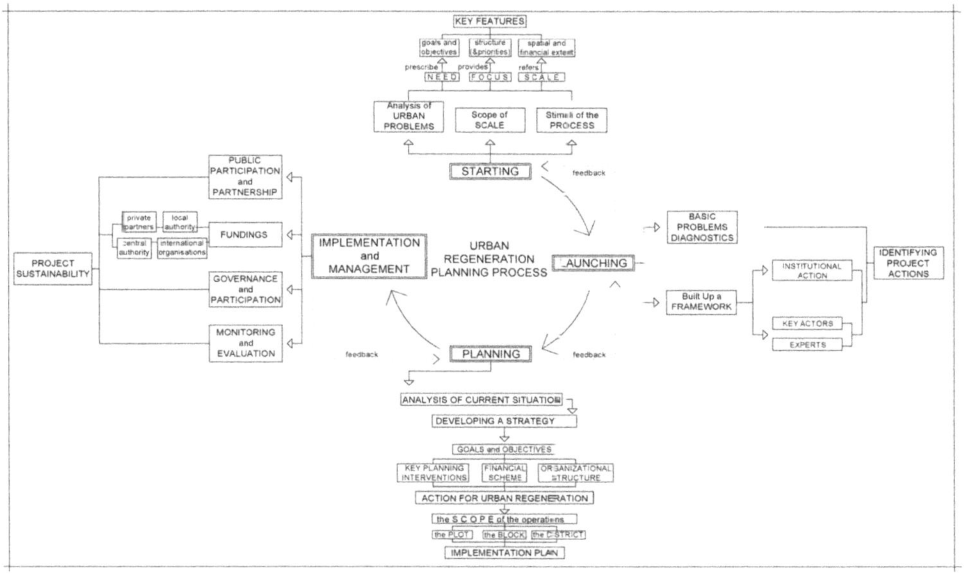


## Historic Centre of Naples, Italy

Naples is one of the most important cities in southern Italy, and it is part of the Mediterranean region. It is the capital of the Campania region, where 3.5 million people live in its metropolitan area, making it the second-largest city in Italy in terms of population (Ahmed, M. et al., 2009). In contrast, the city's surface area is relatively one of the smallest of the large Italian cities, approximately 117.27 km2. Furthermore, population censuses conducted in recent years show a shrinking trend; the 1971 census was 1,226,000, while the 2001 census counted 1,004,500 inhabitants (Mazzeo, 2009). The most recent estimate confirms its population reduction with 973,132 inhabitants (Figure 3.

Dating back to its history, the city was founded by the Greeks in the first millennium BC, making it one of the oldest continuously inhabited urban centres in the world (Blackman & Lentini, 2010). In the 9th century BC, a colony known as Parthenope was established, which became Neapolis in the 6th century BC. It was an important part of Magna Graecia, being central to the merging of Greek and Roman societies, resulting in an important cultural centre for the Romans (Giampaola & Longobardo, 2000). It was the capital of the Duchy of Naples in 661, then became the Kingdom of Naples in 1282, which was annexed to Italy at its unification in 1861 (Musi, 2003). In 1925, Naples was enlarged and modernised by Benito Mussolini's government, only to suffer later severe damage from bombing during World War II (Wanderlingh,

*Figure 3. Naples, Italy (Source: Ahmed, M. et al., 2009)*

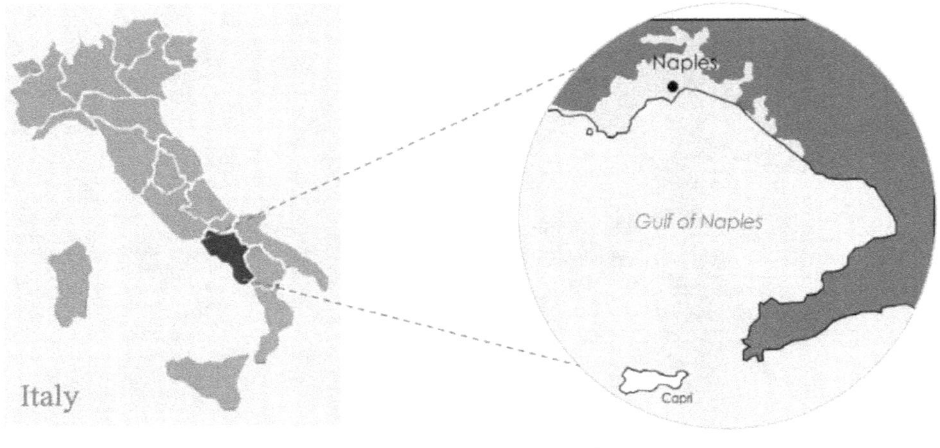


2010). However, after 1945, the city was largely rebuilt, preserving the urban and architectural values of its historic centre.

The historic centre of Naples is one of the largest and oldest in Europe. Due to its outstanding historical value, it was declared a UNESCO World Heritage Site in 1995. The site's significance is given by the permanence of the Greco-Roman urban layout represented in the compact urban fabric composed of buildings dating from 400 to 800. Moreover, the artistic heritage, the handicraft activities, part of the local economy, and the services offered to tourists are part of the intangible heritage resources of the site (Losasso & D'Ambrosio, 2014). Nevertheless, this area has entered a process of continuous physical and social deterioration despite its historical relevance. One of the most notorious elements of this problem is the poor management of solid waste produced by locals and visitors (Chifari et al., 2017). Figure 4 shows the historic centre of Naples.

The current SWM system is based on a traditional scheme of collection trucks taking waste to landfills without any prior waste separation or recycling process. In 2008, decree 4/2008 set a target for the entire southern region to recycle 65% of the waste produced by 2015. While many cities have reached the first target of 40% recycling, Naples has achieved only slightly less than 9%. At present, the city produces 1600 tons of waste every day. With this low recycling rate, almost 67% of solid waste ends up in landfills. Cities that reached the recycling target included in their waste management system technologies such as incinerators, which at the same time generate energy in various ways. In Naples, the first incinerator is still under construction after more than eight years of being commissioned. In addition, there is a high probability of a waste emergency due to the lack of new landfills available,

*Figure 4. The historic centre of Naples (Source: Pane, 1971)*

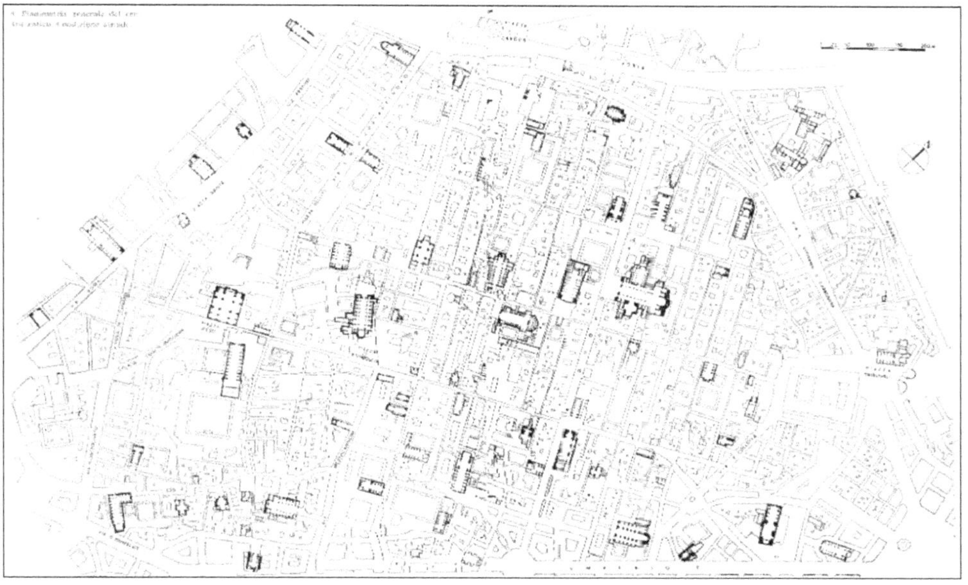

as areas around Naples refused to allow them in their territories due to experiences of mismanagement and hazardous waste in the past (Ahmed et al., 2009).

In the case of the historic centre, it is even more critical (Figure 5). On the one hand, due to the narrowness of the streets, waste collection by trucks is inefficient and unsustainable as it could cause significant damage to the tangible heritage, hampering its historical value (Khan, 2015). On the other hand, the lack of education and awareness campaigns on the relevance of proper waste management, possible recycling and separation at source is non-existent, neglecting the possibility of an improved life cycle for the disposal of recyclables (Chifari et al., 2017). Consequently, it is vital to raise awareness of the significant damage that poor SWM can cause to this historic site. It could lead to finding a regenerative strategy that provides a pathway to contribute to the sustainability of the tangible and intangible heritage. To ensure community involvement in this task, a participatory waste management plan is key to maintaining the built environment. At the same time, locals find an alternative income that contributes to their livelihoods, and visitors are integrated through more sustainable forms of tourism.

*Figure 5. Waste disposal problem in historic centre of Naples (Source: Spiegel Magazine, 2010)*

## Historical Centre of Cairo, Egypt

The Greater Cairo Metropolitan Area, which encompasses Cairo, Giza, Helwan and New Cairo, is the largest urban agglomeration in Africa, with 21.7 million as of 2022, a 2% increase from 2021. Simultaneously, Cairo is considered the largest city in the Arab World, with over 10 million, and has been known as the "City of the Thousand Minarets" due to its distinctive skyline. The population density is 19,376 people per square kilometre, which ranks it the 37th in the world.

Historic centres are layered with the different periodic times that passed through and formed a specific city, and historic Cairo is no different. Cairo transformed and evolved throughout history, starting with the establishment of the city with the first settlements in the early Islamic era (640- 969), then the royal and military refuge during The Fatimids period (969-1171), opening the city to the public and new housing during The Ayyubids period (1171-1250), and the great growth and recovery of the city during The Mamluks (1250- 1516). The main characteristics throughout that period were the mixture of residential areas (including palaces and houses of different sizes), commercial markets, inns (khans), schools, and cultural and recreational venues, used by other social and economic classes over the centuries. As well as the continued use of local materials and traditional building methods.

However, during The Ottoman period (1516-1980) Cairo became a significant economic centre developing new centralities such as The Citadel and Downtown Cairo.

The main features during this time were the change in the social fabric as elite and high to middle-income groups moved out of the historic centre, transforming itself into a shelter for low income and marginalised groups. As a result, the historic centre went from being a place of great relevance to losing its hierarchy and deteriorating.

In 1979, UNESCO declared it a World Heritage site recognizing its "absolutely unquestionable historical, archaeological and urbanistic importance". Moreover, in July 2010, UNESCO-WHC launched the Urban Regeneration Project for Historic Cairo (URHC). This project is included within the larger framework of technically assisting the Egyptian Government in managing the World Heritage Site. Within the framework of the URHC project, it was relevant to stress the importance of keeping the historical significance of the site and its rich heritage alive while considering current challenges such as informal or illegal construction, low access to the services, the inadequate communication between the community and the government, and the lack of a waste management plan (URHC, 2012). Figure 6 shows the boundaries of the historic centre of Cairo.

In 2012, a study on SWM was done as a part of the URHC project to have a database for the aspects and issues related to solid waste and the legislative framework that governs its collection and disposal. According to this study, there is no legislative framework for solid waste in Egypt, and the responsibility for SWM is divided between the Ministry of Environment for the management and its impact on the environment, while the Ministry of Housing, Utilities and Urban planning are in charge of the selection of disposal sites (URHC, 2012). In general, Cairo is estimated to produce 11,000 tons/day of waste, where the household waste generated accounts for 60% while the rest is commercial waste. Only 70% to 80% of that waste is collected, and the rest stays on the streets and open areas.

In contrast, there are several informal collection and recycling activities taking place in urban areas where there is a potential for collecting plastic, glass, or metal. Roamers and scavengers tend to roam the streets picking through the dumpsters, looking for recyclable materials, especially plastics and metals, and eventually trading what they find. Robabikia peddlers are yet another example of informal solid waste collection, where the peddlers go through the streets to collect old items from households such as hardware, which they refurbish and resell in the suburbs of Cairo (URHC, 2012).

Zabbaleen is considered one of the most essential and most extensive communities when it comes to solid waste collection and recycling (96,000 people), as they collect waste door-to-door on variable frequencies. Zabbaleen take the solid waste back to their areas, where they sort and recycle what has been collected. There are two main options when it comes to recycling the solid waste within the Zabbaleen community, the first is the already existing recycling workshops within the Zabbaleen district, and the second is selling the sorted waste to recycling factories outside the area

*Figure 6. The historic centre of Cairo (Source: URHC, 2014)*

or the existing workshops. In 1990, transferring waste by donkey-pulled carts was forbidden, and resources were invested in trucks and licence fees. Until the 2000s, Zabbaleen was the main actor in waste collection (URHC, 2012).

As Historic Cairo is a famous district with a high urban density (which falls within the South and West zones of Cairo governorate), it is one of the highest waste producers in Greater Cairo, mainly due to the commercial and hand-crafted industries. Solid waste management in the South Zone is facilitated by the Cleansing and Beautification Agency (CCBA) and ECARU, while the West Zone is facilitated by the Italian company AMA Arab. Historic Cairo is no different from the rest of the city, as informal waste collection and recycling takes place within the centre, focusing on other materials such as bread, cardboard, textile, and bones (URHC, 2012).

There are two transfer stations within Historic Cairo, where each one receives 1200 and 700 tons of waste per day, and there are no plans to establish new ones due to the limited space. Moreover, the door-to-door is the current collection method as it is a procedure that doesn't require large equipment such as container collection. However, since Historic Cairo is operated by different entities, it leads to shortages in waste collection, which ultimately forces the households to take their garbage to the containers provided in the main streets. Unfortunately, that is not the case, as the locals are not willing to commute there daily (Figure 7). Consequently, the narrow streets end up being filled with garbage (URHC, 2012), a situation that greatly affects the tangible and intangible value of the historic area.

## DISCUSSION AND CONCLUSION

Solid waste management has become one of the most pressing issues on the international agendas of countries in the 21st century, as all seek a more comprehensive way to close the waste cycle. In historical centres, it is even more critical, as poor solid waste management threatens the sustainability of tangible and intangible heritage. Therefore, it is imperative to include a new approach that provides a solution that not only serves to maintain these significant historic areas but also provides a regenerative perspective in which the people who live, work, and visit these places are a fundamental part of it.

In the case of the historic centres of Naples and Cairo, the discourse suggests that there are similarities in current waste management systems. In both cases, separation of waste at source is non-existent, resulting in all waste being placed in the same bag regardless of whether it is organic, recyclable, or ordinary. Furthermore, due to the narrow streets in these areas, the collection system is door-to-door, which is functional but ineffective as its frequency varies, and it is a source of greenhouse gases

*Figure 7. Waste disposal situation in historic centre of Cairo*

due to the trucks that must wait for the collectors. Consequently, as the collection frequency is not daily, this waste remains on the streets for days, affecting the image of the historic centre and the sense of ownership of locals and visitors.

In contrast, while Naples has a formal solid waste collection system, Cairo has both formal and informal collection systems, which adds another layer of complexity to stakeholders' management and coordination, affecting the disposal methods of the waste. In the case of Naples, waste is taken to the landfill without any previous waste processing. This gap is especially alarming because there are no new areas to implement more landfills, which could result in a solid waste emergency. Moreover, in Historic Cairo, there is a gap between the rate of waste produced every day and the collection rate by formal or informal entities. As a result, only 60% of the waste is collected and taken to landfills or to the Zabbleen area, while the rest is left on the streets. Consequently, it directly affects the tangible and intangible heritage of the historic centre, as empty plots are used as dump sites, narrow streets and alleys are contaminated with garbage, which hinders walkability and negatively affects the health of the inhabitants and the visitors.

*Figure 8. The proposed regenerative solid waste management in historical centres*

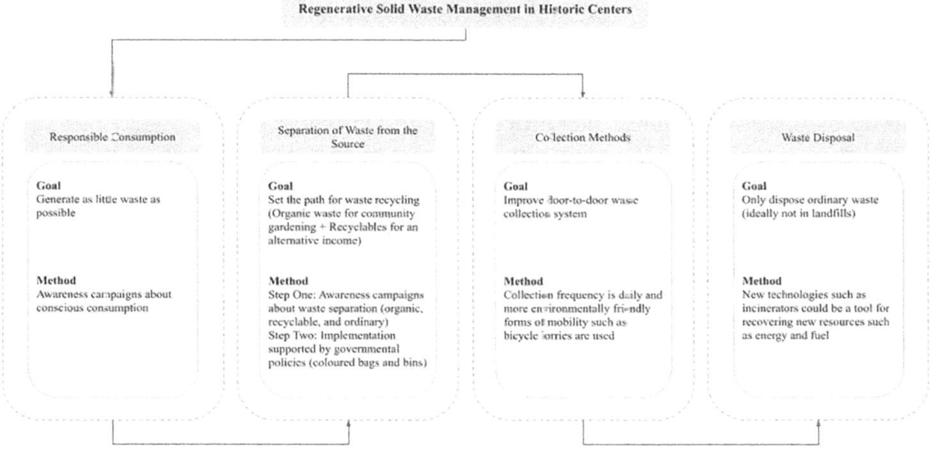


The regenerative approach makes it possible to create holistic strategies that consider economic, socio-cultural, and environmental aspects to produce solutions that go beyond sustainability. The solid waste management system in historic centres can benefit from the different aspects in each cycle phase from these three pillars (Figure 8).

The first phase of the cycle is *Responsible Consumption*. Awareness through campaigns explains the importance of the conscious consumption of products is essential to generate as little waste as possible. The second phase is separating waste from the source, which is divided into two steps. In the first one, awareness is raised among the public about the importance of separating the solid waste into organic, recyclable, and ordinary waste and the potential of using bio-waste as compost. The second one is the implementation. This could be supported by a government policy that includes the promotion of separation through the inclusion of coloured bags and bins in households, commerce, shops, workshops, and streets.

The third phase includes *the* collection methods and is directly related to the waste separation phase. Organic waste could be used as a raw material for community gardening, and the recycling material could generate an alternative income for the collecting entity. While responding to the dense urban fabric, the current door-to-door waste collection system in historical centres could be improved if the collection frequency is daily and more environmentally friendly forms of mobility such as bicycle lorries are used. In this phase, social cohesion is vital in collaboration with restaurants and cafés to collect bio-waste and workshops for recyclable waste.

In the current schemes of both historical centres, the final disposal is in landfills. As a result, the percentage of waste taken to the fourth and final stage, *Waste Disposal,* decreases as only ordinary waste is collected. However, new technologies such as incinerators could be a tool for recovering new resources such as energy and fuel. It is important to note that in the specific contexts of Naples and Cairo, incinerators could be hampered by a lack of political will and corruption.

Consequently, these phases fall directly within the objectives and aspects that urban regeneration focuses on while considering the scale, and the problems of a specific site, as each context has its challenges and potential. The proposed regenerative SWM approach will cover the environmental aspect as dumpsites can be rehabilitated into composting centres and local vegetation plots, which could reduce the pollution rates of the historic centre. Moreover, from an economic aspect, the recyclable waste collected from households and workshops could be used as an alternative income, which in return would expand the local economy. From a socio-cultural aspect, the reduced solid waste on site, which usually hampers the heritage value, will improve the site mobility for locals and visitors, eventually enhancing the user experience and the imageability of the area. The reduced solid waste would enhance the hygiene conditions of the area and become a step closer to improving the overall structure of the area. Finally, with local community participation and their engagement in the community spaces where the vegetation plots are placed, would empower them, and strengthen the social cohesion in the historic centre.

## REFERENCES

Abbas, I. F., Purwaningrum, P., & Indrawati, D. (2021). Evaluation of waste management in tourism area of Luar Batang Village, outside area of the Historic Old Jakarta Kota, Penjaringan District, North Jakarta. *IOP Conference Series. Earth and Environmental Science*, *737*(1), 012012. doi:10.1088/1755-1315/737/1/012012

Ahmed, M., Scerbo, M., Izzo, P., Parrilli, M., Coccia, F., Ganga, V., & Anilir, S. (2009). A Community-based Waste Management System for the Historic Centre of Naples. *Journal of Asian Architecture and Building Engineering*, *8*(2), 363–370. doi:10.3130/jaabe.8.363

Behrens-Abouseif, D. (1989). *Islamic Architecture in Cairo: An Introduction*. E.J. Brill.

Blackman, D.J. & Lentini, M.C. (2010). *Ricoveri per navi militari nei porti del Mediterraneo antico e medievale: atti del Workshop*. Edipuglia srl.

Cairo Population 2022 (Demographics, Maps, Graphs). (n.d.). Available at: https://worldpopulationreview.com/world-cities/cairo-population

Carmona, M. (2021). *Public Places-Urban Spaces: The Dimensions of Urban Design* (3rd ed.). Routledge. doi:10.4324/9781315158457

Carter, J. G., Cavan, G., Connelly, A., Guy, S., Handley, J., & Kazmierczak, A. (2015). Climate change and the city: Building capacity for urban adaptation. *Progress in Planning*, *95*, 1–66. doi:10.1016/j.progress.2013.08.001

Chaline, C., & Coccossis, H. (2004). *Guidelines for urban regeneration in the Mediterranean region. UNEP, Priority Actions Programme*. Regional Activity Centre.

Chifari, R., Renner, A., Lo Piano, S., Ripa, M., Bukkens, S. G. F., & Giampietro, M. (2017). Development of a municipal solid waste management decision support tool for Naples, Italy. *Journal of Cleaner Production*, *161*, 1032–1043. doi:10.1016/j.jclepro.2017.06.074

Ciudin, R., Isarie, C., Cioca, L., Petrescu, V., Nederita, V., & Ranieri, E. (2014). Vacuum waste collection system for an historical city centre. Academic Press.

Darinka Czischke, C. M. (2015). *Sustainable regeneration in urban areas*. URBACT.

Ellen MacArthur Foundation. (2017). *Cities in the Circular Economy: An Initial Exploration*. Available online:https://www.ellenmacarthurfoundation.org/publications/cities-in-the-circular-economy-an-initial-exploration

Fahmi, W., & Sutton, K. (2010). Cairo's Contested Garbage: Sustainable Solid Waste Management and the Zabaleen's Right to the City. *Sustainability*, *2*(6), 1765–1783. doi:10.3390u2061765

Farahat, B. I. (2020). Employing sustainable development concepts to revitalize the historic urban quarter of Al-Muizz Street, Cairo, Egypt. Academic Press.

Fayed, L., Elshater, A., & Rashed, R. (2020). Aspects of Regenerative Cities. In *Architecture and Urbanism: A Smart Outlook* (pp. 303–319). Springer. doi:10.1007/978-3-030-52584-2_22

Gaber, D. A. (2019). *From Waste Management to Waste Markets: Infrastructure*. Industry and Business Opportunities.

Gidarakos, E., Havas, G., & Ntzamilis, P. (2006). Municipal solid waste composition determination supporting the integrated solid waste management system in the island of Crete. *Waste Management (New York, N.Y.)*, *26*(6), 668–679. doi:10.1016/j.wasman.2005.07.018 PMID:16207528

Girardet, H. (2014). *Creating regenerative cities*. Routledge. doi:10.4324/9781315764375

Hornsby, C., Ripa, M., Vassillo, C., & Ulgiati, S. (2017). A roadmap towards integrated assessment and participatory strategies in support of decision-making processes. The case of urban waste management. *Journal of Cleaner Production*, *142*, 157–172. doi:10.1016/j.jclepro.2016.06.189

Battuta, I. (2009). The Travels of Ibn Battuta (S. Lee, Trans.). Cosimo Classics. (Original publication 1829)

Italian Agency for Environment Protection and Technical Services. (2007). *The Waste Report 2007*. Author.

Khan, S. (2015). Revitalizing historic areas; Lessons from the renovation of Saidpur Village, Islamabad. *Journal of Research in Architecture and Planning*, *18*(1), 11–22. doi:10.53700/jrap1812015_2

Kaza, S., Yao, L. C., Bhada-Tata, P., & Van Woerden, F. (2018). *What a Waste 2.0: A Global Snapshot of Solid Waste Management to 2050*. World Bank. doi:10.1596/978-1-4648-1329-0

Lang, T. (2005). *Insights in the British debate about urban decline and urban regeneration* (No. 32). Working Paper.

Laso, J., García-Herrero, I., Margallo, M., Bala, A., Fullana-i-Palmer, P., Irabien, A., & Aldaco, R. (2019). LCA-Based Comparison of Two Organic Fraction Municipal Solid Waste Collection Systems in Historical Centres in Spain. *Energies*, *12*(7), 1407. doi:10.3390/en12071407

Losasso, M., & D'Ambrosio, V. (2014). Progetto ambientale e riqualificazione dello spazio pubblico: Il grande progetto per il centro storico di Napoli sito Unesco. *Techne. Journal of Technology for Architecture and Environment*, *7*, 64–74.

Mazzeo, G. (2009). Naples. *Cities (London, England)*, *26*(6), 363–376. doi:10.1016/j.cities.2009.06.001

Musi, A (2003). Napoli, una capitale e il suo regno. *Touring, 118*, 156.

Nathanson, J. A. (n.d.). Solid-waste management. *Encyclopedia Britannica*. https://www.britannica.com/technology/solid-waste-management

Tanrıkul, A., & Hoşkara, Ş. (2019). A new framework for the regeneration process of Mediterranean historic city centres. *Sustainability*, *11*(16), 4483. doi:10.3390u11164483

Punkkinen, H., Merta, E., Teerioja, N., Moliis, K., & Kuvaja, E. (2012). Environmental sustainability comparison of a hypothetical pneumatic waste collection system and a door-to-door system. *Waste Management (New York, N.Y.)*, *32*(10), 1775–1781. doi:10.1016/j.wasman.2012.05.003 PMID:22647502

Raymond, A. (1993). *Le Caire*. Fayard.

Roberts, P. (2000). The evolution, definition and purpose of urban regeneration. Urban Regeneration, 9-36.

Rose, C., & Boxberger, L. (1995). Ottoman Cairo. In Cairo: Living Past, Living Future. The University of Texas Centre for Middle Eastern Studies.

Soares Henriques, L., John, N., Harseim, L., Ababei, P., & Mujica von Seggern, M. J. (2017). *Climate Responsive Urban Planning in El-Gamaleya*. Decentralized Greywater Recycling.

Swelim, T. (2015). *Ibn Tulun: His Lost City and Great Mosque*. American University in Cairo Press.

Wanderlingh, A. (2010). Naples: The History. Intra Moenia.

Waste Management in Cairo: An Informal Success. (2019). *The Borgen Project*. Available at: https://borgenproject.org/waste-management-in-cairo/

Ma. (2015). Sustainable water systems for the city of tomorrow—A conceptual framework. *Sustainability*.

# Chapter 5
# Blue-Green Infrastructure for Heat Exposure Mitigation:
## Alexandria, Egypt

**Riham Nady Faragallah**
*Pharos University in Alexandria, Egypt*

## ABSTRACT

*Global and regional warming phenomena drive the effect of climate change on cities. Research on urban heat adaptation strategies has been growing, with many studies concentrating on blue and green infrastructure (BGI) as an alternative to grey infrastructure and to minimise the adverse effects of urban heat islands. Building upon investigating the case of Germany, this chapter presents mitigation strategies for Alexandria, Egypt that can reduce the temperature at the urban design level and identify urban heat island effects through BGI and their performance, as well as their importance in achieving sustainable urban development and low-temperature cities. The lessons learned show that BGI is a new approach to city planning that can minimise the impact of human development on functioning natural systems, the quality of life, and the economic consequences. A set of BGI guidelines for Alexandria on how to keep cities cool and improve the city's environment were agreed upon.*

## INTRODUCTION

Rapid changes in global temperature trends and climate have resulted due to the increase of greenhouse gases in the Earth's atmosphere. These changes are predicted to have significant safety, security and health effects, in addition to disrupting daily lives and causing environmental losses in urban areas. Another recent phenomenon

DOI: 10.4018/978-1-6684-2462-9.ch005

that has exacerbated this issue is the Urban Heat Island (UHI) effect, where a metropolitan area can be on average 1.0-6.0$^{0}$C warmer than nearby non-urban areas (Dimoudi, Kantzioura, Zoras, Pallas and Kosmopoulos, 2013).

According to research, cities currently account for 60-80% of the world's overall energy usage (Chaoui, 2009). It was also mentioned that UHI is a crucial factor that affects heating and cooling loads in buildings, and thus urbanization has a significant impact on energy consumption. Since it has been estimated that the ratio of the world's urban population to the total population is expected to rise from 54% in 2016 to 60% in 2030 (United Nations, 2016). Therefore, UHI is expected to become a severe problem in the future of urban life.

The natural land cover is converted to built surfaces that trap the incoming solar radiation during the day and then re-radiate it at night. As a result, urbanization demand created a decline in vegetated areas and increased the built-up surfaces. Consequently, there has been a severe loss of green spaces to an artificial environment. For this reason, there have been an increase in study on urban heat (UH) adaptation measures, with many pieces of research focusing on blue and green infrastructure (BGI) as an alternative method to grey infrastructure (GI) and a tool to reduce the negative impacts of urban heat island (UHI).

At the local level, blue-green infrastructure has been highlighted as a "best practice" for generating greater resilience and sustainability. It is increasingly recognized and valued as a solution for addressing the urban water and green challenges posed by current climate variability and urbanization and reinforcing the cities to future change. In addition, the implementation of BGI is driven by the need to minimize urban heat and improve thermal comfort in towns. Environmental concerns in urban planning are not a new concept, and they have played a leading role in the evolution of planning theory and practice. Still, the idea of blue-green infrastructure is relatively new.

Humans cause not only environmental problems in Alexandria city, Egypt, but climate change poses significant threats, mainly heat stress in urban areas and an increase in urban heat islands, both of which have a considerable influence on the city's inhabitants. This is due to the city's high-density buildings and streets accommodating a large population. Urban expansion threatens land supply, causes traffic volume growth, increases pressure on the environment, and makes the city unsustainable. Thus, like many other cities, Alexandria faces the problem of climate change, the force of scarcity of resources, high pollution and severe environmental degradation, which cause dangers to human health. Consequently, the lack of reaction to these issues will have disastrous problems.

These problems were observed along the Mahmoudiya Canal (west of Alexandria city), a central concern in all Alexandria's Governorate plans for the past 40 years. There has never been a plan or strategy that consider microclimate conditions that directly affect human comfort along the canal. During the new developments and projects in the last two years, the existing channel and water bodies are either being filled up or lost due to excessive pollution. Green spaces were also being destroyed. Accordingly, there is an urgent need to act on several influences that have reached critical levels, including high levels of UHI, loss of thermal comfort, increase in greenhouse gases and lack of water bodies and biodiversity.

This chapter hypothesizes that the idea of BGI may have a beneficial impact on UHI to examine possible applications of blue and green infrastructure as a part of achieving the low-temperature city concept and creating sustainable urban development. The reason is that a low-temperature city has sustainable solutions and adaptation methods for urban heat management. As a result, this chapter assumes that BGI techniques and methods may positively contribute to cooling cities and enhancing urban thermal comfort.

The chapter is divided into three main parts to reach its goals. The first part explores the physical changes in cities due to rising temperatures, particularly the UHI phenomenon. Also, it presents an overview of UHI, its adverse effects on urban areas and methods that manage urban heat problems by which municipal governments adapt to change urban climates through the concept of BGI. In addition, the chapter presents the BGI, its principles, benefits and potentials. The chapter examines how BGI successfully controls temperatures and minimizes energy usage effectively and assesses its demonstrated effectiveness in regulating urban temperatures. The second part analyses the example of Stuttgart, Germany, that applied BGI strategies to reduce UHI and the practical measures used to create low-temperature cities. The last and third part examines the current situation along the Mahmoudiya Canal and the existing problems associated with UHI. This part ends with a plan of BGI for decreasing the adverse effects of UHI and the best practices and lessons gained from recent BGI programs in climate change adaptation measures, urban water management and biodiversity promotion. An example is given to deteriorating vegetation conditions, lack of water bodies and increase in urban heat island, all of which neglect the thermal comfort of residents and visitors to this metropolitan area. Then, this section deals with the ways of adapting UHI through BGI presenting new approaches to city planning that attempt to minimize the impact of human development on ecological systems, the quality of life and the economic consequences. Finally, this leads to propose a set of guidelines and principles for low-temperature cities and enhancing the urban environment in Alexandria generally and along the Mahmoudiya Canal, mainly by applying BGI methods (Figure 1).

The specific aim of the study is to consider the microclimate conditions along the Mahmoudiya Canal, as well as lower the urban level's temperature and identify the UHI mitigation strategies through BGI. Furthermore, it aims to decide legislative measures to enhance thermal comfort, achieve sustainable urban development and promote low-temperature cities.

*Figure 1. Methodology flowchart*
*Source: (The author)*

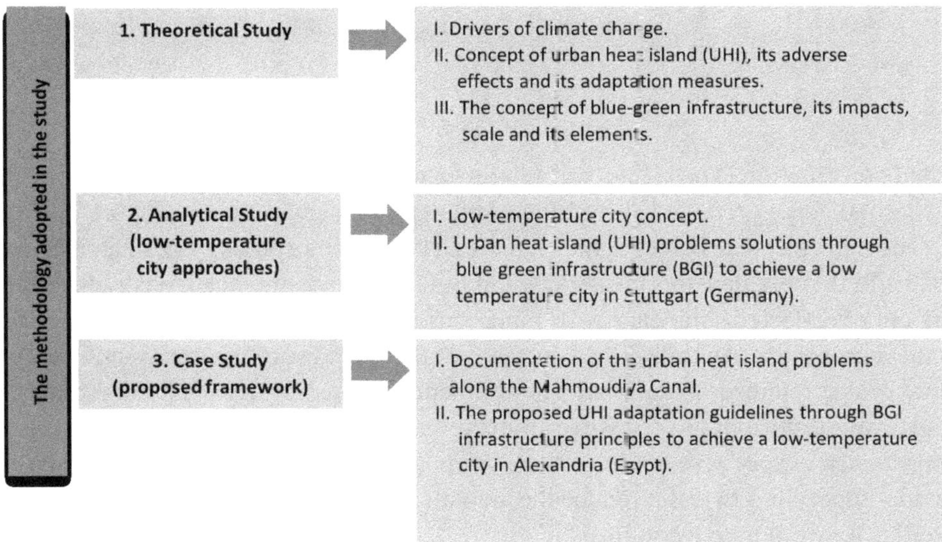


# URBAN SCALE CLIMATE CHANGE

## Drivers of Climate Change in Cities

Global and regional warming phenomena drive climate change in cities. On the one hand, rising greenhouse gases concentrations on a global scale enhance the natural greenhouse effect, which traps outgoing radiations from the Earth's surface and warms the atmosphere. On the other hand, at the regional scale of cities, four distinct changes in urban environments give rise to different warming mechanisms. In recent decades, the urban heat island effect (UHI), the study's primary focus, has been discovered to be the leading cause of rising temperatures in cities. Also, these physical changes in cities include the high density of urban morphology that traps solar radiation and the emission of waste heat from buildings and vehicles. The introduction of non-vegetative surface materials is more efficient at absorbing and storing thermal energy than natural land covers and the loss of natural vegetation to

*Figure 2. City warming mechanisms*
*Source: (The author)*

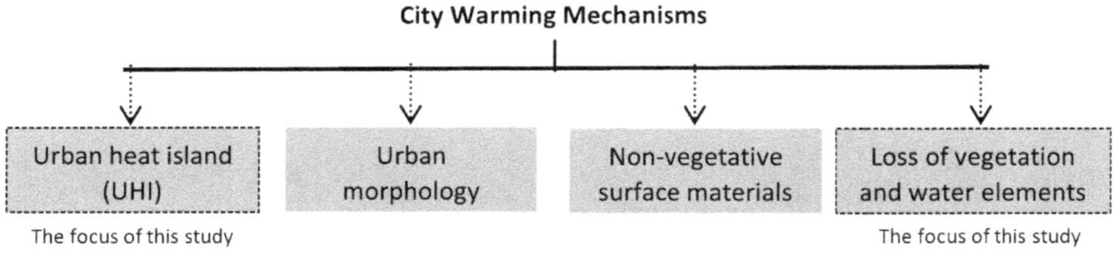


urban construction. These four warming mechanisms in cities elevate the quantity of thermal energy retained and emitted into the urban environment (Figure 2).

The main physical driver of the UHI effect is urban morphology (Elshater, et al., 2022). In densely developed districts, tall buildings and street canyons reduce the amount of reflected solar energy from the surface that may pass unobstructed back to the atmosphere. As the vertical surfaces of the city absorb this reflected energy, more heat is retained in the urban environment. The loss of trees and other natural land covers contribute to a warmer climate through decreasing shading and, most importantly, evaporative cooling. Also, urban construction materials such as street paving impervious to water reduce the amount of moisture absorbed and contained in cities for evaporative cooling.

Finally, cities are zones of high energy consumption due to vehicle usage, building cooling and heating loads and industrial activities. Waste heat is produced as massive amounts of energy are consumed in urban environments, and it is eventually vented to the atmosphere leading to rising temperatures (Climate Urban Lab, 2016).

## Overview of Urban Heat Island Effect (UHI)

Urban heat island (UHI) has drawn considerable attention in recent years and is becoming a critical issue worldwide as cities rapidly develop (Figure 3). The term "urban heat island" (UHI) refers to a phenomenon in which the temperature in urban areas is higher than that in rural areas (Balany, Ng, and Wong, 2020). During the daytime, urban structures absorb solar heat (radiation) and release it back to the environment at night-time (Oke, 1981).

The heat islands (HI) happen in many layers or areas of the urban environment (Voogt, 2004). This effect is caused by the lack of natural evaporative surfaces (vegetation), the physical characteristics of the ground cover such as concrete and asphalt, which absorb rather than reflect solar radiation and other activities, urban geometries (barriers to relieve heat and slow winds), air pollution and energy use

*Figure 3. The urban heat island phenomenon*
*Source: (The author)*

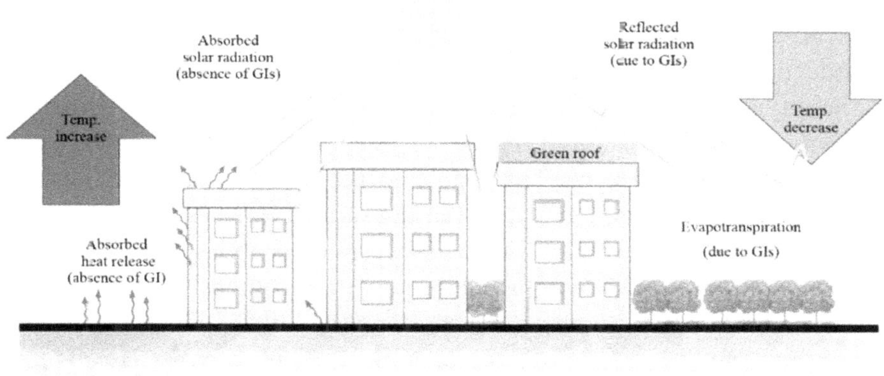


(Filho, Wolf, and Castro-Diaz, 2020). The UHI is a multi-faceted phenomenon. There are several types of UHI (*boundary*, *canopy* and *surface* layer heat island), each describing a different urban heat (UH). In addition, (Voogt 2004) explained: "the increasing temperature of the urban air settings refers to the boundary and canopy layer heat islands".

The urban canopy layer (UCL) encompasses the urban cover layer below the average height of buildings. People's experience of the climate and high temperatures occurs mainly at the micro-scale, in the layer of air between the ground and the top of roofs or trees, known as the "urban canopy layer" (UCL). The "classic" UHI effect refers to changing air temperature with UCL. This UHI is most noticeable on warm, still, summer nights, when the inner urban air temperature might be 3.5-4.5C° warmer than the surrounding rural areas. It is solid in densely populated areas.

As urbanization proceeds, the need for accurate weather forecasting at the urban scale becomes critical (Musco, 2016). The urban boundary layer (UBL) is the part of the atmosphere and one of the most complex and least-understood climates. The urban boundary layer (UBL) encompasses the urban cover layer above the average height of buildings (Barlow, 2014). The surface heat island refers to the temperature over the entire three-dimensional surface of a city, e.g. roofs, walls, and street surfaces. The surface UHI is affected by solar radiations, so the most significant temperature differences are most pronounced during the day, particularly in open areas exposed to sunlight. Solar radiation has also been one of the most critical elements in determining human thermal comfort under hot conditions (Filho, Wolf, and Castro-Diaz, 2020).

The immediate surrounding environment influences the climate experienced at the micro-scale (e.g., street environment), the local-scale environment (e.g., neighbourhood environment) and the city-wide environment (e.g., synoptic influences and UHI) (Norton, Bosomworth and Coutts, 2013). The city-wide UHI combines heat production and retention at the micro and local scales.

As urban heat island produces significant changes in the local microclimate, the intensity of UHI has manifold consequences at the regional scale, which can be clustered under different aspects. This chapter focuses on the reduction of high urban surface temperatures. Consequently, decreasing surface temperatures will lower air temperatures during day and night. Also, it focuses on adapting urban heat islands at the micro and local scale environment through BGI methods such as (vegetation and water bodies). Modifications of the environment at these scales will influence people's experience of the climate within UCL (city-wide atmosphere). They can have flow-on effects on larger-scale climatic processes, including reducing the UHI if implemented over large areas (Figure 4).

*Figure 4. Levels of adapting urban heat island*
*Source: (The author)*

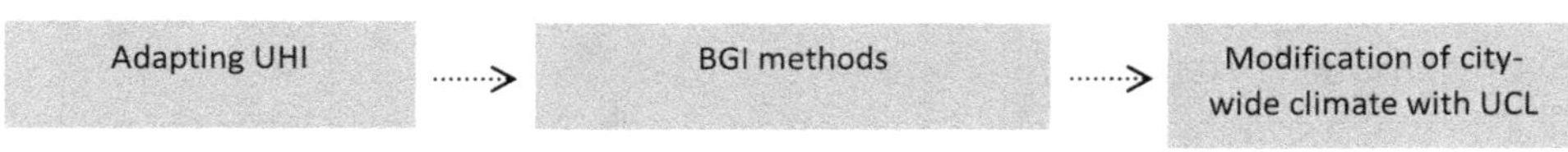


## Adverse Effects of Urban Heat Island

UHI has extensive effects on the urban environment and infrastructure that support society. The intensification of an urban heat island has a regional-scale impact on air quality, energy demand and public health. UHI phenomenon raises the energy demand, speeding up the development of hazardous pollution and causing human thermal discomfort and health problems by increasing heatwaves over metropolises (Figure 5).

Summertime UHI drastically affects the outdoor air quality and increases the city's energy demand. Due to this energy intensification, extensive power blackout may occur due to increased air conditioning usage. Reports show that thousands of people die annually due to heat-related diseases. In addition, to the impact of electricity and temperature consumption, UHI also increases pollution concentration over cities. Furthermore, it affects the local climate by changing wind patterns, forming fog and clouds, increasing humidity and altering the perception rate (Mirzaei and Haghighat, 2010).

*Figure 5. Impacts of UHI on urban environments*
*Source: (The author)*

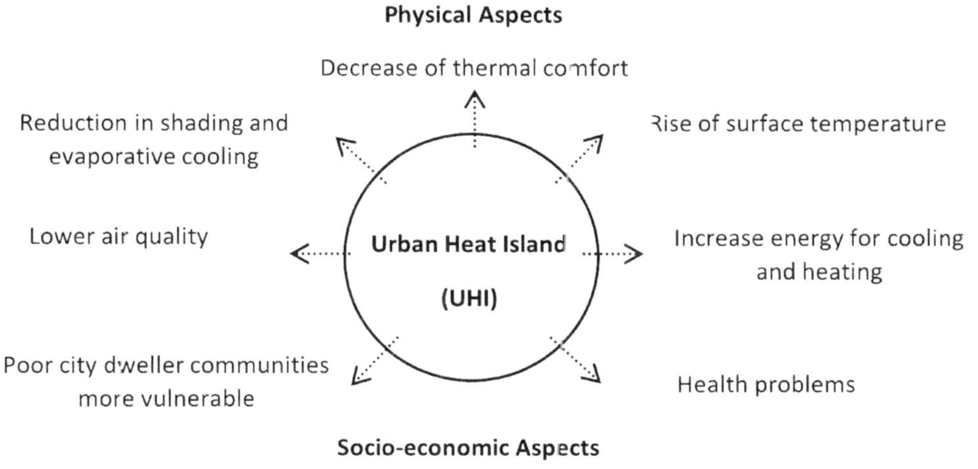


Cities with poor climatic conditions consume more power to cool buildings in summer and at the same time for lighting. Moreover, residents frequently experience discomfort and difficulty due to high temperatures (Arabi, Shahidan and Kamal, 2015). As cities increase in size, more land area is covered with impervious surfaces that trap and retain heat and leave less space for green places, reducing UHI and improving ecosystem services. Trees are one form of green space required by many cities to establish and implement comprehensive plans that have a genuine impact (Westendorff, 2021).

## Synergies between Adaptation and Mitigation Strategies of UHI

Neutralizing the impacts of UHI is a primary concern for societies. Because of the problem severity, extensive research has been conducted, and an enormous scope of literature is available for the subject. The literature in this field comprises the concepts, methodologies, latest investigation tools, research approaches and mitigation strategies (Norton, Bosomworth and Coutts, 2013). Since cities are the primary cause of global warming and the emission of greenhouse gases, heat island adaptation is an approach for minimizing the vulnerability against present and future impacts of climate change rather than the mitigation measures (Parry, Canziani, Patulikof, and Hanson, 2007). However, there can be synergies between mitigation and adaptation in the urban mechanism. Urban adaptation has a more significant potential for comprehensive efforts involving several sectors. This adaptation can also take many forms (e.g., structural, non-structural), where structural urban adaptation involves new techniques and implementations and non-structural adaption involves communication and management plans (Carmin, Anguelovski, and Roberts, 2012).

To make cities more resilient to climate change issues, there are opportunities for planners and designers to consciously deal with green and blue spaces in the urban landscape (Demuzere, Orru, Heidrich, 2014). Environmental considerations in urban planning are not a new idea, and they have played a leading role in the development of planning theory and practice. Still, specifically, the concept of blue-green infrastructure (BGI) is comparatively new. In this context, the blue-green infrastructural approach can be considered a tool for delivering ecosystem services in environmental planning policies. As a result, adaptation can focus on the importance of ecological, social, and economic factors (Nahyan, 2017). In this chapter, the potentials of BGI for transformation are targeted for specific local climate change impacts.

## BLUE-GREEN INFRASTRUCTURE (BGI)

The topic of blue-green infrastructure has become a well-established concept in urban environmental planning, policy, research, and design, with growing awareness and understanding of its potential benefits for ecology and society (Wouters, Dresiseitl, and Wanschura, Unknown). The term blue-green infrastructure refers to urban water bodies, vegetation and greenery in cities, which have several additional benefits for quality of life in cities (environmental, social and economic benefits) in the form of ecosystem services in addition to their ecological functions (Machac, Louda and Dubova, 2016). Blue-green infrastructure solutions are characterized as "planned and managed natural and semi-natural systems that can provide more categories of advantages, compared to traditional grey infrastructure. It aims to succeed in the sustainable resilient of nature's ecosystems (Abdou and Tarek, 2016).

A blue-green infrastructure strategy describes an urban area planned and designed to incorporate natural systems that provide the ecological and amenity value associated with urban greening and stormwater management. Often "blue" assets (Water Sensitive Urban Design – WSUD, drainage areas and flood storage) and "green" assets (trees, parks, gardens) are planned separately. However, a single investment may often provide multiple services that help with both "blue" and "green" objectives (Department of Environment, Land, Water and Planning, 2017).

The term "blue infrastructure" technically refers to "hydrological infrastructure including rainwater and urban stormwater systems as well as surface water and groundwater aquifers". In urban design, blue infrastructure is traditionally considered a matter of resilient water supply and security provisions. Such water infrastructure may be natural, adopted or artificial and provided functions of slowing down,

*Figure 6. The concept of blue-green infrastructure (BGI)*
*Source: (The author)*

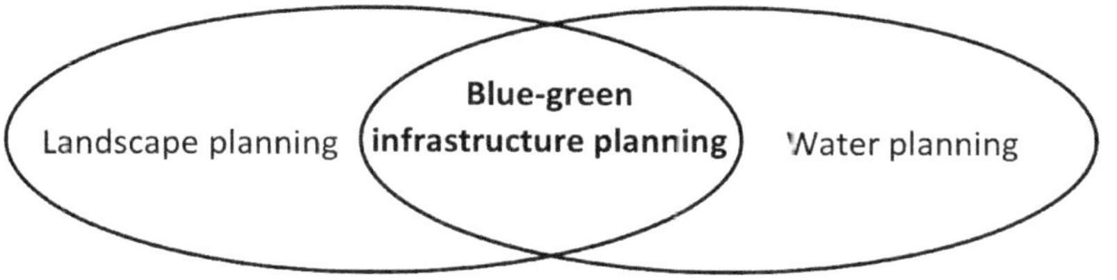


decentralization and spreading, soaking into the underground, evaporating and releasing water into the natural environment. This includes flow control, detention, retention, filtration, infiltration and different forms of water treatment like reuse and recycling. In general, blue infrastructure addresses aspects of water quantity and quality control.

The term "green infrastructure" refers to projects that include vegetated design elements such as parks, green roofs, green belts, alleys, vertical and horizontal gardens and planters. However, the term "green infrastructure" is a bit of a misnomer, as these infrastructures are often closely associated with "blue" processes (Figure 6). The BGI paradigm combines these two types of infrastructures and value in a union that is greater than the sum of its parts (Demuzere, Orru, Heidrich, 2014).

## Impacts and Potentials of BGI

The impact of BGI is acknowledged in environmental, social and economic effects. Concerning the ecological effects, blue-green infrastructure reduces fuel consumption. It minimizes air, water and soil pollution by using natural elements and techniques such as green roofs, green walls, street trees and non-vehicular travel routes such as walking and cycling, reducing heat-induced. Connected green infrastructure assets may be used to establish green corridors more resilient biodiversity to face the change of climatic effects. This network also connects existing urban projects and new landscape areas using linear blue elements such as streams and canals within natural features such as parks and linear routes.

Meanwhile, the green infrastructure recognizes the social effects of green areas within the urban building blocks. It may serve as a focal point for community participation through public involvement in designing, delivering, and managing new assets and providing education, training, and volunteering opportunities. On the one hand, green infrastructure (GI) reduces health inequalities, improves mental well-being by providing access to nature and attractive green spaces, and encourages physical activities by creating quality green spaces for walking, cycling, sports and playing, reinforcing the local landscape character adding a solid identity. On the other hand, blue infrastructure can provide open spaces for people to gather along rivers, canals, lakes, ... etc., which enhances social interaction between users and inhabitants of different ages. On the same line of thinking, economic impacts are reflected by the blue-green elements are available with affordable cost as they are local natural components of the urbanized local areas (Nahyan, 2017).

### Scale of BGI

Successful blue-green infrastructure (BGI) planning naturally requires an integration of different scales of planning and application. Accordingly, blue-green infrastructure design will likely be for a whole urban area but might need to consider possible interventions at all scales and their cumulative impact (Abdou and Tarek, 2016). BGI should be regarded at all planning scales, from the strategic framework to the neighbourhood and streets' level and down the building. At the site level, covering rooftops and wall surfaces with green and vegetation as they cover a large built surface area.

Promoting innovative and appropriate planting in street design at the street level helps soften the street scene by creating visual interest and improving the microclimate. Neighbourhood level, integrating existing roads, paths and surrounding development, creating opportunities for easy and suitable access into the green network and avoiding bad sitting for buildings which could block the entrance to the green network.

At the strategic level, thinking of how development can join up to the existing green network by creating plugs to the gaps in the strategic green network and landscape setting and benefiting the new product by providing good connections and a continuous web of routes (Nahyan, 2017). Table 1 shows the elements of BGI.

## URBAN HEAT MANAGEMENT THROUGH LOW-TEMPERATURE CITY APPROACHES

SDG11 is a spatially organized development framework that seeks to address several interrelated issues by focusing on urban sustainability. It aims to achieve the desired outcome level by 2030 through a set of attainable targets and indicators designed to guide policy actions at local levels. Like other SDG goals and targets, it is based on the system approach, emphasising cross-linkages with other developmental priorities. Thus, Goal 11 is most relevant and seeks to make cities and human settlements inclusive, safe, resilient and sustainable through eliminating slum-like conditions, reducing urban sprawl, addressing urban resilience and climate change challenges,

*Table 1. Scale and features of BGI*

| Scale | BGI Element | Description and Characteristics |
|---|---|---|
| Site Level | Green roofs | A green roof is a building roof that has been partially or entirely covered in vegetation and is planted into a growing medium sitting above a waterproof membrane. Harvested rainwater can be used for irrigation. |
| | Green walls | A green wall is a vertical garden on the side of a building that comprises vegetation planted within a growing medium and is attached to the wall. |
| Street Level | Street trees / Vegetation | Trees planted in a growing medium underneath the sidewalk can be designed to be passively irrigated from stormwater runoff from pavements and roads. These can also enhance stormwater pollutant removal by including special filter media. |
| | Stormwater | Permeable paving can also channel stormwater into underground soil areas to support trees. |
| City Level | Gardens | Gardens comprise vegetation planted into a growing media (soils). Stormwater can be directed into gardens to provide passive irrigation, or an active irrigation system can be supplied and fed by alternative water sources. |
| | Urban parks and open green spaces | Parks are public open space areas that provide the local community with recreational activities. They are most typically characterized by a mix of turf, shrubs and trees which can yield critical cooling benefits to the urban environment. These could be irrigated using an alternative water supply or designed to provide stormwater detention and infiltration. |
| | Urban rain gardens | Raingardens are garden beds that are designed to capture, detain and treat stormwater runoff as it filters through the underlying media before it is discharged at the base of the system either into the surrounding soils or into the local stormwater network. |
| | Swales | Swales are shallow, vegetated open channels that convey and treat stormwater. |
| | Sports grounds | Sports grounds are large open space areas that support active recreation activities. These could be irrigated using an alternative water supply or designed to provide stormwater detention and infiltration. |
| | Green corridors | Green corridors are linear green spaces that provide connectivity services, including natural habitat and recreational pathways. These areas are typically located along with waterways or other easements. |
| | Ponds and lakes | Ponds and lakes are open water bodies that are designed to hold water permanently. Vegetation can be included around the edge or in shallow sections. A stormwater supply or recycled water can feed them. |
| | Canals | Canals are natural or artificial waterway channels for water conveyance or servicing water transport vehicles. Canals are used to transport water for irrigation and other human uses. |
| Strategic Level | Urban agriculture | Urban agriculture is the local production of food products. This can include community gardens open to the public or commercially viable small-scale urban farms. |
| | Waterways | Waterways are channels that capture and convey flows from catchments. They include streams and creeks and can be natural or modified systems (e.g., rock-edged or even concrete-lined). |
| Neighbourhood Level | Forests | Forests are large areas of dense trees, shrubs and ground covers. Forests play an essential part in the water cycle, creating a porous space to absorb stormwater. |
| | Wetlands | Wetlands are heavily vegetated water bodies. These systems can either be natural features in the landscape or constructed to treat stormwater. They can appear as biological systems or integrated as hard-edged features in urban areas. |

Source: (The author)

better management of urban environments (pollution and waste management), providing access to safe and secure public spaces for all and improving urban management through better urban policies and regulations (Chatterji, 2020).

As a result, this chapter introduces a low-temperature city as a sustainable urban solution for the future of cities that depend on urban heat management. It is a critical factor for decreasing urban heat emissions, creating solutions for future climate change by reducing global emissions and creating smart growth and low-temperature community scenarios.

There are many objectives of a low-temperature city on urban planning and design. It promotes blue infrastructure to improve water supply, manage water scarcity and contribute towards a more stable climatic future. Also, it encourages the use of green infrastructure and green areas to support low-temperature in urban environments to reduce UHI through trees and vegetation that directly shade buildings and decrease demand for air conditioning to improve air quality and lower greenhouse gas emissions to achieve environmental sustainability.

## Low-temperature City Applications: The Case of Stuttgart City

The UHI project aims at developing mitigation and risk prevention and management strategies concerning the Urban Heat Island (UHI) phenomenon. The UHI project is being developed in eight of the most significant metropolitan areas and Mega Urban Regions of the Central Europe Cooperation programme. This activity is supported by the Central European Programme and the European Regional Development Fund. Accordingly, this section will focus on the example of Stuttgart City as it has adopted several sustainable solutions and management approaches for UHI to solve the environmental problems it was facing through applying the concept of the low-temperature city. This section aims at concluding the adaptation measures of UHI at the level of urban design that could be used in the Egyptian context.

## Location and Climatic Conditions

Stuttgart is a well-known city for its industrial productivity and affluence and is often referred to as Germany's "cradle of the automobile". It is characterized by its dense, sociable city centre and its lush surrounding hillsides. Stuttgart is located in the southwest region of Germany in a valley basin, and it is considered the lowest-temperature city in the world. Stuttgart has been planned to respect and protect nature and exploit how natural wind patterns and dense vegetation can assist the city in combating overheating and air pollution problems. Stuttgart's climate planning strategy is regarded as one of the best examples of heat island management on a global scale.

Stuttgart's mild climate, low wind speeds, industrial activity, and high traffic volume have made the city highly vulnerable to poor air quality. The development of the valley slopes has prevented them from moving through the city, worsening air quality and contributing to the urban heat island effect (Kapp, 2021). A Climate Atlas was developed for the Stuttgart region, presenting the temperature and cold air flows distribution according to the city's topography and land use. Based on this information, several planning and zoning regulations were recommended to preserve environmental quality, manage climatic conditions, and increase open spaces in densely built-up areas (Department of Environment, Land, Water and Planning, 2017).

## Urban Challenges

Since the city is located in the valley basin, it cannot quickly get rid of its heat, staying trapped in the town. According to some studies, climate projections suggest that the number of days with heat stress will increase significantly until 2050, owing to more frequent heatwaves. The built-up areas of the city of Stuttgart could experience more than 60 days of heat stress by 2050. It is expected that the number of days with heat stress will continue to rise till the end of the century (Department of Environment, Land, Water and Planning, 2017).

More than 60% of the city is covered with vegetation to face this problem. Vineyards, forests and public parks surround the city, extending down the hillside to the city's centre. This unique landscape feature and the particular topography of the region are employed to create an attractive and comfortable urban environment for Stuttgart's residents (Rehan, 2016).

To further improve resilience to hotter summers and more frequent heatwaves, different adaptation green measures have been implemented in the city, including green roofs, greening of tram tracks and buildings, shading facades of buildings through street trees and upgrading small public spaces to "cool spots". Ventilation corridors and expanded green areas also improved urban air quality, dispersing and reducing air pollutants (Department of Environment, Land, Water and Planning, 2017).

Also, numerous blue infrastructure approaches have been improved and applied, such as drinking fountains and other water elements.

All these adaptation measures helped to reduce the heat island effect that will further increase due to hotter summers and more frequent and intensive heat waves.

## Combating the Heat Island Effect through Blue-Green Infrastructure

The correct application of blue-green infrastructure (BGI) can combat the urban heat island effect. In Stuttgart's case, this is made part of urban policy in such a way that supports other macro layers which offer environmental benefits, increase bio-diversity and promote air quality. A new approach for the strategic planning of blue-green infrastructure investments was directed towards improving essential linkages between greenspaces and places of employment and living. These critical linkages, particularly for commuters within Stuttgart Region, are the "development axes" mainly along the river valley like Neckar, the Rems and the Fils (Urban Climate Lab, 2016). There are many examples of blue-green infrastructure applications as follows:

1. **Green corridors:** Stuttgart is a forerunner in protecting green spaces by creating green ventilation corridors and banning construction strategically. It has not only covered its climate with winds that hinder overheating, but it has also improved air quality and increased resilience against global warming. With the support of detailed local climate maps, Stuttgart has stopped planned construction (Jacobson, 2012). At night clean air sweeps down from the surrounding hills and runs through a series of "ventilation-corridors" which have been kept open as wide, tree-flanked arteries within the city's street infrastructure.
2. **Urban parks:** Stuttgart has become one of Germany's greenest metropolitan cities by focusing on the city's sustainable development based on the principle of fantastic city. More than 60% of Stuttgart city is a green area. There are 65,000 trees in parks and 35,000 trees along streets throughout the city. Since the 1970s, the city has integrated green spaces into a giant green "U", an urban massive public park. It is possible to go through park surroundings from the central royal gardens to the forests at the city's edge. Among the park planners were ecologists and social scientists who aimed at a more ecological approach for the Rosenstein Park, encouraged contact between people and nature. Additionally, the park made the city cooler through green areas. Besides that, Stuttgart has a lot of parks such as Mansion Mountain Park, the Wilhelma, the Leibfried Garden, the Waiting Mountain and the Killesberg Elevator Park (Urban Climate Lab, 2016).
3. **Vegetation/compact urban green:** Stuttgart's land-use plan is urban-compact-green which is an excellent example in the project of Stuttgart 21 and is a sustainable vision in urban design and planning due to its objectives of making the city cooler through maintaining the continuity of Rosenstein Park by transforming the present rail tracks into green urban spaces. On the other

hand, most of the trees in Stuttgart are scarcely planted, sometimes trees have been planted just on one side of the street, sometimes not at all, or with a lot of space in between. Thus, compact urban green has a vital role in diminishing urban heat emissions (Urban Climate Lab, 2016).

4. **Green roofs:** Stuttgart is one of the leading cities in the world in adopting green roofs. Since 1993, all new buildings in Stuttgart have been greened, and it has subsidized more than 3000,000 square meters of green roofs. Although green roofs are expensive to construct, heavy with deep soil profiles, and require much maintenance, they have many benefits. They are mainly installed for environmental performance, visual improvement and air cleaning. The green roofs can maintain cooler indoor temperatures. The apartment under the top of the building will no longer feel the summer heat, perhaps most importantly, eliminating the greenhouse effect (Grig and Chris, 2011).
5. **River path:** The Neckar River cycling path is essential for commuters and tourists in the Stuttgart region. It is considered one of the green infrastructure investments in the city. The new route sections allow green, safe and efficient transportation between industrial and commercial centres residential and leisure areas in the Stuttgart region. In addition, the cycle route provides a way to include exercise in the daily life of individuals and improves the greening and landscape quality along the river (Urban Climate Lab, 2016).
6. **Urban water:** Water can reduce temperatures by evaporation, through transporting heat out of the city and as a buffer that slowly absorbs the heat. Water applications, in general, are more effective when it is moving or dispersed as done by a fountain. Cooling by water evaporation depends on the air movement that distributes the cool air across the city (Kleerekoper, 2009). Consequently, Stuttgart depends on fountains in the urban area, and numerous lakes are visible along Rosenstein Park (green U Park).

## CASE STUDY

This section deals with the Mahmoudiya Canal to document the most critical factors in increasing UHI and the problems that face its local urban environment. Following that, the chapter suggests a framework for urban heat management through BGI and a set of future recommendations to create a low-temperature city in Alexandria based on the case study results to reduce the UHI and make the Mahmoudiya Canal a more relaxed urban area.

*Figure 7. The Mahmoudiya Canal*
*Source: (The author)*

## Location and Climatic Conditions

Alexandria is the second-largest city and metropolitan area in Egypt after Greater Cairo in size and population. It extends about 32km along the coast of the Mediterranean Sea on the northern border. The Mahmoudiya Canal passes by its south linking the River Nile to the Mediterranean Sea (El Darwish, Ragheb, and Sherif, 2015). Alexandria city is characterized by a hot climate which is generally humid. The temperatures are hot or very hot on summer days and warm or mild on winter days. Also, it has high humidity levels in summer due to its location by the Mediterranean Sea (Figure 7).

Alexandria City is experiencing problems that prevent making it a low-temperature city. Many factors contribute to raising the UHI in Alexandria city generally and along the Mahmoudiya Canal specifically, such as the following (Figure 8):

- The high density of the population is due to the rapid growth of urbanization.
- Climate change causes heat stress in urban areas and consequently an increase in UHI.
- The environmental pollution due to increased vehicles, traffic congestion and unsustainable urban transportation systems.
- The lack and decline of vegetation.
- The scarcity of water resources.
- The loss of land covers and impervious street paving materials.
- High industrial activities lead to rising temperatures.

## Documentation of Urban Issues Along El Mahmoudiya Canal

First, at the neighbourhood level, the focus of the study area is along the Mahmoudiya Canal. There are a lot of informal settlements along the Mahmoudiya Canal that

*Figure 8. The current problems along the Mahmoudiya Canal*
*Source: (The author)*

were built without planning and have little or no planned infrastructure or green spaces. Furthermore, these informal settlements feature crowded buildings that raise the UHI effect. Second, the Mahoudiya Canal passes through Alexandria city at the urban level, where several industrial uses contribute to raising the UHI. For example, several industries (El Nehas and El Karnak Paper industries, Alexandria Company for Metal Products and Alexandria Company for Spinning and Weaving, ...etc.) prevent air movement and increase pollution. Along the Mahmoudiya Canal, a certain number of parks exist, such as Antoniadis Park and Alexandria Zoo, however, these represent a small percentage of the area of the Mahmoudiya and the area is experiencing a decrease in the presence of green spaces that influence air quality.

Furthermore, the Mahmoudiya Canal lacks green corridors, which raise the UHI. Another problem is that most urban water bodies along the Mahmoudiya Canal are filled- up or losing depth due to excessive pollution. Thirdly, there is an absence of applications towards sustainability at the level of buildings, street networks and streetscape. For example, there is a lack of green roofs, and green walls and the materials used in the streetscape are unsustainable. This is clear in the pavement and is made worse by the absence of landscaping along the sidewalks and street corners, leading to heat absorption and negatively affecting the environmental aspects and urban image (Figure 9).

*Figure 9. The causes of urban heat island effect along the Mahmoudiya Canal*
*Source: (The author)*

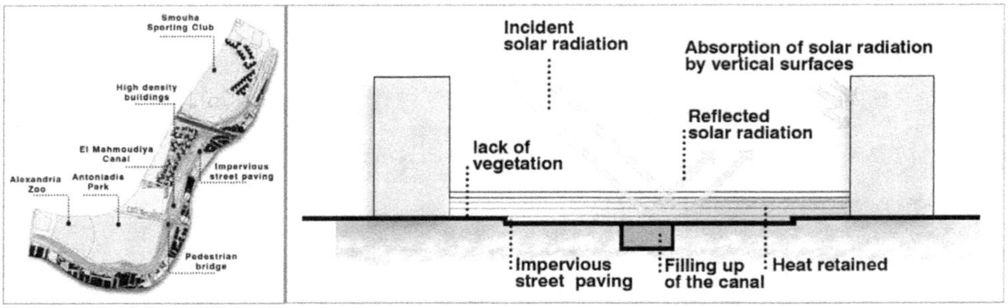


## Urban Heat Island Adaptation Plan through BGI Applications (Solutions)

Accordingly, this section focuses on BGI applications along the Mahmoudiya Canal as a way to solve heat stress problems and decrease UHI in the area and transform Alexandria into a low-temperature city at site, street and city level (local-micro climate) (Figure10).

### Albedo

1. **Low-temperature buildings:**
    - Encourage the use of reflective materials and natural cooling systems in buildings.
    - Construct several zero-energy buildings which depend on natural ventilation and require no energy for cooling and heating.
    - Improve buildings that use passive ventilation to reduce the dependence on air-conditioning.
2. **Reflective materials:**
    - Encourage the use of high-albedo pavements to reduce the area's ambient temperature.
    - Change the colour of pavements, roofs and facades by using light colours.

### Green Infrastructure

1. **Urban Parks:**

- Increase the number of urban parks by creating a large central park along the Mahmoudiya Canal or establishing an urban park network (connect Alexandria Zoo with Antoniadis Park).
- Large trees should be planted to provide shading and mitigate UHI.

2. **Green corridors:**
   - Transform the Mahmoudiya Canal axis into a green corridor to improve the urban environmental quality and promote the city's image, making it more attractive and competitive to inhabitants and visitors.
   - Improve air quality by using green ventilation corridors and minimizing heat gains in urban environments.
3. **Street trees:**
   - Provide shade for pedestrians through large trees and planting on both sidewalks of El Mahmoudiya Canal.
   - Encourage the installation of landscaped sidewalks.
4. **Green roofs and walls:**
   - Promote green roofs to enhance air quality and contribute to the aesthetic integration of buildings in the landscape.
   - Encourage green walls to improve the energy efficiency of buildings.

## Blue Infrastructure

1. **Water installations:**
   - Encourage the integration of water installations into the landscape to enhance the urban design, minimize environmental degradation, and reduce UHI.
   - To generate cooling, water should be integrated along the Mahmoudiya Canal and in parks such as fountains.
   - Conserve water resources through reuse and system efficiency to improve the urban climate.
2. **El Mahmoudiya Canal cycle path:**
   - Establish a route along the Mahmoudiya Canal for pedestrians and cyclists and provide this route with a sustainable streetscape that includes solar lighting, vegetation, sustainable materials in floor finishing to reflect the sun's radiation.
3. **Drainage of rainwater:**
   - Manage rainwater through, for example, effective runoffs and street drainage systems.
   - Side street gully traps drain rainwater through the surface water system to recycle and reuse rainwater for other purposes.

*Figure 10. Proposed UHI adaptation solutions along the Mahmoudiya Canal*
*Source: (The author)*

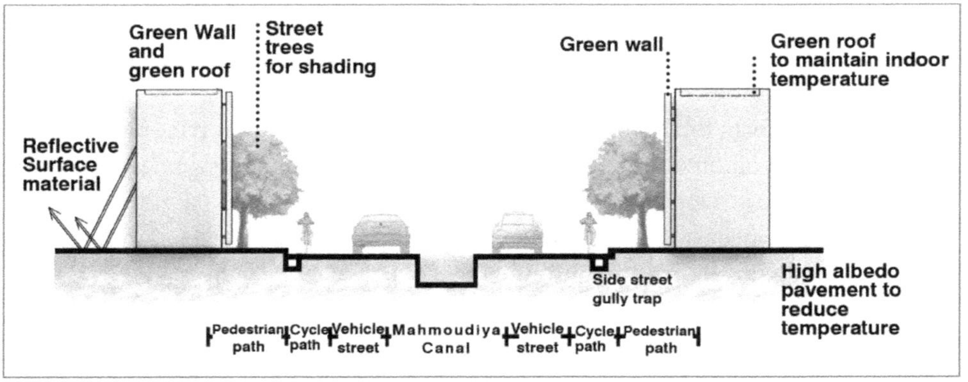


## COMPARISON BETWEEN STUTTGART CITY AND THE MAHMOUDIYA CANAL

After analysing the causes of UHI in both Stuttgart city and along the Mahmoudiya Canal, the following section discusses the adaptation measures between both cases that will reduce the accumulation of heat in urban areas. This emphasizes that cities can mitigate their UHI effect by planning actions, which are considered the main strategies for low-temperature cities. These strategies can be summarized as the following: green corridors, vegetation, urban water, green roof and walls, … etc.

## A FRAMEWORK FOR URBAN HEAT ISLAND MANAGEMENT

This section discusses the measures in the form of a framework that will reduce the accumulation of heat stress in urban areas. Cities can adapt to the UHI effect by planning actions and adaptation approaches, the main strategies to cool the city (Figure 11).

The framework is proposed to manage urban heat islands along the Mahmoudiya Canal, offer an opportunity to achieve a low-temperature city, and describe the modifications that reduce heat stress. The design strategy of a low-temperature city is to minimize indoor temperature in buildings that require no energy for cooling and promote passive techniques for natural ventilation. Also, it encourages the use of reflective surface covering (light colour) to reduce the ambient temperature. Furthermore, it enhances the establishment of urban parks that help to minimise UHI. This requires a group of specialists in urban planning and design. Asides

*Table 2. Analysis between Stuttgart city and the Mahmoudiaya canal and the applications for urban heat management*

| Point of Comparison | Stuttgart City (Germany) | El Mahmoudiya Canal (Alexandria, Egypt) |
|---|---|---|
| **1. Location** | Southwest region of Germany in a valley basin | West of Alexandria |
| **2. Climatic conditions** | A climate Atlas was developed for cold air flows according to topography and land use | Hot climate generally humid |
| **3. Urban problems** | High industrial activity<br>High traffic volume<br>Poor air quality<br>High UHI | Informal settlements with high density.<br>Industrial activities<br>Decrease in green spaces<br>Lack of green roofs and walls<br>Unsustainable ground covers<br>Water bodies are filled-up |
| **4. Blue-green infrastructure (BGI) solutions** | | |
| **(a) Green corridors** | Green corridors for vegetation through using tree flanked arteries | Using large trees to shade pedestrian paths |
| **(b) Urban parks** | Integrating green areas through the large "U" urban public park | Creating large central parks |
| **(c) Vegetation** | Transforming rail track into green urban spaces | Transform large areas into green spaces |
| **(d) River path** | A cycling path along the river with green and efficient transportation | A cycling path along the canal to promote daily activities |
| **(e) Urban water** | The use of fountains and numerous lakes to transport heat out of the city | The use of effective drainage systems and side street gully traps |
| **(f) Green roofs** | Adapting green roofs to maintain cool indoor temperatures | Adapting green roofs and green walls to maintain indoor temperatures |

Source: (The author)

from that, the framework suggests using green roofs and walls to keep the buildings cool on summer days and cut the need for energy-intensive air conditioning. This framework offers to convert some streets to green corridors, which improve the urban environment quality and to create a cycle path along the Mahmoudiya Canal with vegetation.

These planning actions may create a linked system that supports urban life while respecting the natural environment. In addition, it provides proper application of blue-green infrastructure, which can help mitigate the UHI effect and lead to management measures concerning this phenomenon.

In this way, the city will become a social-environmental settlement that can have its system in which the urban growth is accepted and managed based on environmental considerations (Masnavi, 2011). This may be included in urban policy so that it complements other macro layers that offer environmental benefits, increasing the quality of life and decreasing harmful effects. Adaptation and coordination of BGI and UHI are inevitable for the city's sustainability, and it is on behalf of the city to adapt itself to the environmental systems.

## DISCUSSION

There is a strong correlation between UHI effects and BGI elements to achieve a low-temperature city. This was clear after analysing the case of Stuttgart city and the Mahmoudiya Canal in terms of solving the problems of heat stress and BGI strategies and the planning actions to achieve a low-temperature city (Figure 12).

It was clear that street plants and shading trees provide environmental benefits by filtering pollutants reducing the noise level, and resulting in numerous economic benefits. Several studies indicated significant summer energy saving by providing shade trees next to buildings. Trees modify the microclimate by altering the solar radiation from the ground by providing shading. Shading is critical in reducing summer air temperatures and mitigating UHI (Wong, 2005). Therefore, increasing the area covered by trees will bring beneficial effects in the shaded areas and lead to lower air temperatures compared to non-shaded areas (Sawka, 2013).

Relevant literature indicated that the application of vegetation in urban areas would alter the microclimate parameters such as air temperature, relative humidity, wind pattern and precipitation. Therefore, urban greening has always been recommended as an essential adaption strategy and practical approach to mitigate the heat island phenomenon and reduce the related consequences of increased air temperatures (Gill, Handley, Ennos and Pauleit, 2007). However, large urban parks have significant cooling effects, extending towards their surrounding environment. Urban gardens have a great potential to mitigate the heat island effect and improve pedestrian thermal comfort (Oliveira, 2011).

According to the literature review, high albedo pavements contribute to an improved climate of the cities and reduce the surface temperature (Jamei, 2016). In addition, pavement material absorbs solar radiation and dissipates part of the accumulated heat to the atmosphere, resulting in high air temperature. Also, a review of studies on the effect of bicycle paths on the outdoor environment showed the beneficial effect of these paths on cooling cities. Although increasing the number of water installations was effective in providing low-temperature urban areas. However, the location of these installations concerning the city has to be considered. Several studies have been conducted to ensure the impact of rivers and canal paths (water elements) on pedestrian level thermal comfort. All these links help achieve low-temperature community strategies, which promise to reduce energy use and improve the liveability of cities by mitigating urban heat islands.

*Figure 11. A proposed framework for urban heat management strategies and planning actions for low-temperature cities*
*Source: (The author)*

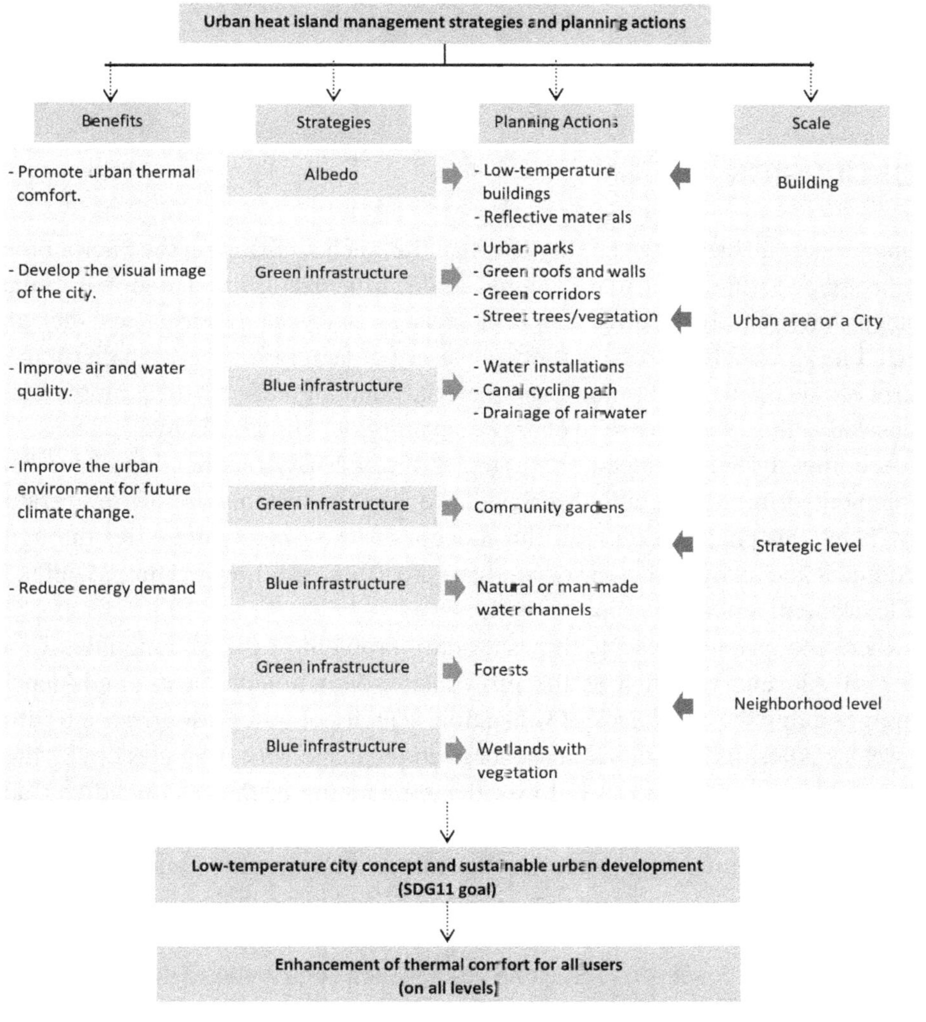

*Figure 12. Future recommendations to adapt UHI*
*Source: (The author)*

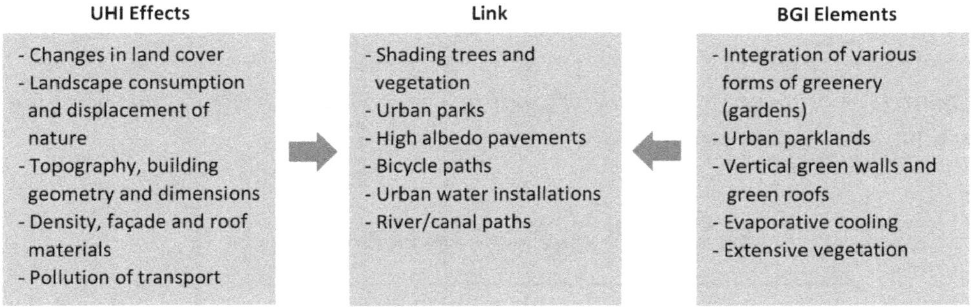


## CONCLUSION

Many aspects influence the quality of urban life, such as changing the microclimate of the metropolitan area. On a global scale, urbanization and regional climate changes have rapidly occurred in recent decades. In these circumstances, the value of UHI has probably increased due to a variety of factors such as high distribution morphology, the height of buildings, materials and human activities, more traffic, industrial activities, increase in energy consumption, climate change, … etc.

Accordingly, UHI adaptation strategies through BGI applications have been implemented in various cities worldwide to reach a sustainable low-temperature city. Thus, this chapter examines the concept of low-temperature cities through a theoretical and analytical study of Stuttgart city (Germany), which turned out to be the lowest temperature city in the world.

Then, the chapter presents the case study problem in the Mahmoudiya Canal (west of Alexandria), such as the growth of informal settlements, high-density buildings along the canal, lack of vegetation, scarcity of water resources, and filling up water bodies. Its adaptation approaches and management strategies could direct urban designers and planners to lower the temperature of the Mahmoudiya Canal and reduce the UHI effect. The study also proposes using reflective materials in buildings and pavements to minimize the ambient temperature. Finally, the chapter concludes the main elements for reducing the UHI, which includes the BGI strategies identifying opportunities for urban greening, making buildings more energy-efficient and following the concept of urban corridors, which contributes to the improvement of urban environmental quality.

Finally, the chapter suggests future actions to deal with the UHI effect along the Mahmoudiya Canal. It proposes a framework through which Alexandria could be a low-temperature city, specifying the modifications to reduce the UHI effect.

If implemented systematically, all these strategies will stand as a good chance for reducing the negative impacts of UHIs in the built urban environment.

From a broad perspective, it could be concluded that the principle of urban design that affects UHI and microclimate needs to be considered through several recommendations based on the above study. The supposed remarks figured out the essential need to use albedo surface materials to reduce the mean radiant temperature. Furthermore, the results highlight the importance of choosing simulation tools for reliable results before urban development. These two findings open future research to stand on other parameters such as shading elements or building orientation that affect the emergence of UHI effects.

## REFERENCES

Abdou, A., Abd Elgawad, I., & Tarek, T. (2016, July). Green Infrastructure to achieve Sustainability in Urban Design: Nile Corridor in Great Cairo as a Case Study. *Journal of Al Azhar University of Engineering Sector.*, *11*(40), 1055–1066. doi:10.21608/auej.2016.19410

Arabi, R., Shahidan, M., Kamal, M., Jaafar, M., & Rakhshandehroo, M. (2015). Mitigating Urban Heat Island Through Green Roofs. *Current World Environment*, *10*(1), 918–927. doi:10.12944/CWE.10.Special-Issue1.111

Balany, F., Ng, A., & Wong, M. (2020, December) Green Infrastructure as an Urban Heat Island Mitigation Strategy- A Review. *Water (Basel)*, *12*(3577), 3577. Advance online publication. doi:10.3390/w12123577

Barlow, J. (2014). Progress in observing and modelling the urban boundary layer. *Urban Climate*, *10*, 216–240. doi:10.1016/j.uclim.2014.03.011

Carmin, J., Anguelovski, I., & Roberts, D. (2012). Urban Climate Adaptation in the global South Planning in an Emerging Policy Domain. *Journal of Planning Education and Research*, *31*(1), 18–32. doi:10.1177/0739456X11430951

Chaoui, K. (2009). *Competitive cities and climate change*. OECD Publishing.

Chatterji, T. (2020). *SDG 11 Sustainable Cities and Communities: SDG 11 and the New Urban Agenda: Global Sustainability Frameworks for Local Action*. Springer Nature. doi:10.1007/978-981-32-9927-6_12

Climate Urban Lab. (2016). *The benefits of green infrastructure for heat mitigation and emissions reductions in cities*. The Trust for Public Land.

Demuzere, M., Orru, K., Heidrich, O., Olazabal, E., Geneletti, D., Orru, H., Bhave, A. G., Mittal, N., Feliu, E., & Faehnle, M. (2014). Mitigating and Adapting to Climate Change: Multi-functional and Multiscale assessment of Green Urban Infrastructure. *Journal of Environmental Management, 146*, 107–115. doi:10.1016/j.jenvman.2014.07.025 PMID:25163601

Department of Environment, Land, Water and Planning. (2017). *Planning a Blue-Green City: A how-to guide for planning urban greening and enhanced storm water management in Victoria*. E2design Lab.

Dimoudi, A., Kantzioura, A., Zoras, S., Pallas, C., & Kosmopoulos, P. (2013). Investigation of urban microclimate parameters in an urban center. *Energy and Building*, *64*, 1–9. doi:10.1016/j.enbuild.2013.04.014

El Darwish, I., Ragheb, A., & Sherif, A. (2015). The Role of Simulation in Urban Design Decisions: Microclimate and Human Comfort Considerations in Planning. *Architecture and Planning Journal, 23*(1).

Elshater, A., Abusaada, H., Alfiky, A., El-Bardisy, N., Elmarakby, E., & Grant, S. (2022). Workers' satisfaction vis-à-vis environmental and socio-morphological aspects for sustainability and decent work. *Sustainability*, *14*(3), 1699. doi:10.3390u14031699

Filho, W., Wolf, F., & Castro-Diaz, R. (2020, January). Addressing the Urban Heat Island Effect: A Cross-Country Assessment of the Role of Green Infrastructure. *Sustainability*, *13*(753), 753. Advance online publication. doi:10.3390u13020753

Gill, S., Handley, J., Ennos, A., & Pauleit, S. (2007). Adapting cities for climate change: The role of the green infrastructure. *Built Environment*, *33*(1), 115–133. doi:10.2148/benv.33.1.115

Grig & Chris. (2011). *The New Green Spaces in Cities-The Green Roofs*. http://www.howtobuildahouseblog.com/new-green-spaces-in-cities-green-roofs

Jacobson, M. (2012). *Stuttgart green corridors*. http://wwf.panda.org/?204461/Stuttgart

Jamei, E., Rajagopalan, P., & Mahmoudian, M. (2016). Review on the impact of urban geometry and pedestrian level greening on outdoor thermal comfort. *Renewable & Sustainable Energy Reviews*, *54*, 1002–1017. doi:10.1016/j.rser.2015.10.104

Kapp, R. (2021). *Stuttgart: Combating the heat island effect and poor air quality with ventilation corridors and green-blue infrastructure*. Climate ADAPI. www.climate-adapt.eea.europa.eu

Kleerekoper, L. (2009). *Urban Heat: Design Principles for Urban Heat Management in the Netherlands* [Master Thesis]. Faculty of Architecture. Delft University of Technology.

Machac, J., Louda, J., & Dubova, L. (2016). Green and Blue Infrastructure: An Opportunity for Smart Cities? *Smart Cities Symposium*, Prague. 10.1109/SCSP.2016.7501030

Masnavi, M. R. (2011). *Sustainable urban forms planning and design strategies: the compact city, urban sprawl and mixed-use development in theory and practice*. LAP Lambert Academic Publishing.

Mirzaei, P. A., & Haghighat, F. (2010). Approaches to study urban heat island – abilities and limitations. *Building and Environment, 45*(10), 2192–2201. doi:10.1016/j.buildenv.2010.04.001

Musco, F. (2016). *Counteracting Urban Heat Island Effects in a Global Climate Change Scenario*. doi:10.1007/978-3-319-10425-6

Nahyan, H. M. (2017). *Planning Green-Blue Infrastructure for Urban Climate Adaptation: A Study Related to Urban Flooding Issues in Dhaka* [Master Thesis]. Land Use Planning, Wageningen University, The Netherlands.

Norton, B., Bosomworth, K., & Coutts, A. (2013). *Planning for a Cooler Future: Green Infrastructure to Reduce Urban heat*. Climate Adaptation for Decision-Makers.

Oke, T. R. (1981). Canyon geometry and the nocturnal urban heat island: Comparison of scale model and field observations. *International Journal of Climatology, 1*(3), 237–254. doi:10.1002/joc.3370010304

Oliveira, S., Andrade, H., & Vaz, T. (2011, November). The cooling effect of green spaces as a contribution to the mitigation of urban heat island: A case study in Lisbon. *Building and Environment, 46*(11), 2186–2194. doi:10.1016/j.buildenv.2011.04.034

Parry, M. I., Canziani, O. F., Patulikof, J. P., & Hanson, C. E. (2007). *Impacts, adaptation and vulnerability. Contribution of working group 11 to the Fourth Assessment Report of the Ingovernmental panel on Climate Change*. Cambridge University Press.

Rehan, R. (2016). Cool city as a sustainable example of heat island management: Case study of the coolest city in the world. *Housing and Building National Research Center Journal, 12*(2), 191–204. doi:10.1016/j.hbrcj.2014.10.002

Sawka, M., Millward, A., Mckay, J., & Sarkovich, M. (2013, May). Growing summer energy conservation through residential tree planting. *Landscape and Urban Planning*, *13*, 1–9. doi:10.1016/j.landurbplan.2013.01.006

Voogt, G. A. (2004). *Urban Heat Islands: Hotter Cities*. Actionbioscience E-Newletter.

Westendorff, V. (2021). *Urban Trees: Strategies for Reducing Urban Heat Island in Cities*. The Field. ASLA Professional Practice Networks' blog. www.thefeild.asla.org

Wong, N., & Yu, C. (2005, September). Study of green areas and urban heat island in a tropical city. *Habitat International*, *29*(3), 547–558. doi:10.1016/j.habitatint.2004.04.008

Wouters, P., Dresiseitl, H., & Wanschura, B. (n.d.). Blue-Green Infrastructure as Tools for the Management of Urban Development and the Effects of Climate Change. *Ramboll Environ.*

United Nations. (2016). The World's [Department of Economic and Social Affairs, Population Division.]. *Cities (London, England)*, •••, 2016.

Urban Climate Lab. (2016). *The benefits of green infrastructure for heat mitigation and emissions reductions in cities: A review of the literature*. The Trust for Public Land's Climate-Smart Cities Program.

# Chapter 6
# Greenery Measures to Mitigate Urban Heat Island in Unplanned Areas:
## Imbaba, Giza, Egypt

**Parisa Kloss**
*Resilient Urban Planning and Development, Germany*

**Heba Allah Essam E. Khalil**
https://orcid.org/0000-0002-8994-6286
*Cairo University, Egypt*

**Aynaz Lotfata**
*Geography Department, Chicago State University, Chicago, USA*

## ABSTRACT

*This chapter applies a design-based approach to integrate the innovative greenery measures in a high-density unplanned area of Imbaba to mitigate the urban heat island (UHI) effect and provide high quality open spaces by creating a chain of cooling public spaces called "Takeeba." The literature review was applied to propagate the formulation of the research problems and identify the boundaries of knowledge. Land surface temperature (LST), the normalized difference vegetation index (NDVI), and the normalized difference built-up index (NDBI) were mapped using remote sensing to identify the hotspots and the existence of greenery and vacant lands for intervention. Prioritizing the hotspots shows vacant land and parking lots are the hottest spots due to direct exposure to solar radiation and heavily compacted areas in the centre of Imbaba. To tackle UHI in the whole Greater Cairo, the replicability of the idea was considered in a follow-up project.*

DOI: 10.4018/978-1-6684-2462-9.ch006

## INTRODUCTION

Rapid urban growth increases concern for the future of cities and the health of inhabitants. It causes the natural environment to be replaced with built-up areas to accommodate overpopulation, quick construction without considering the local climatic conditions, the expansion of city size, high traffic congestion, production of anthropogenic heat, and many others emerged urban microclimate change called urban heat island (UHI). This phenomenon occurs, in fact, by absorbing solar radiation through urban surfaces and storing it during the day; and re-emitting it to the environment at night-time. This creates higher temperature differences between urban and rural areas and leads to several negative impacts on various aspects of life, including increased morbidity and mortality risks, low air quality, and reduction of natural resources. However, health risks are more severe than the other potential threats.

Therefore, the paper focuses on integrating greenery into a compact, unplanned area in Greater Cairo called Imbaba by designing a chain of cooling public spaces in different scales to provide urban thermal comfort for the inhabitants, especially who are more vulnerable to climate change and have less access to essential services and infrastructure as well as to increase their health condition, last but not least, to boost rehabilitation along the Nile and re-establish high quality urban life. It has been proven that greenery is able to decrease the health effects of the UHI by removing pollutants from the air (Solecki et al., 2005) and blocking solar radiation by providing shadow.

Through a mapping exercise of one of the most vulnerable typologies of Greater Cairo, the paper identifies possible assets and challenges in terms of available land/surfaces, water/reused wastewater, and the acceptability of interventions. Then, it suggests a matrix of possible innovative greenery interventions responsive to these different urban structure typologies.

The preliminary analysis done by mapping LST, NDVI and NDBI ascertain that the extent and severity of UHI depend on the sorts of land uses located in the city. Prioritizing the hotspots (high surface temperature) shows that vacant lands and parking lots are the hottest and greeneries are the most incredible spots. It is undeniable that greenery plays a crucial role in regulating surface temperatures. The heavily compact built-up areas are the second hotspots due to the lack of vegetation and using low albedo materials.

The paper proposes designing several interconnected cooling public spaces called "Takeeba" for the identified hottest spots in Imbaba, starting from a vacant land along the Nile, followed by further interventions into the target area and extended into alternative vacant lands near Imbaba park.

The overall goal of the paper is a collaborative contribution to Greater Cairo towards becoming a sustainable city as defined in the New Urban Agenda (NUA) and the United Nations Sustainable Development Goals (SDG) – especially concerning three goals:

- **SDG 3:** 'good health and well-being.'
- **SDG 11:** 'sustainable cities and communities
- **SDG 13:** 'climate action.'

To achieve the goal, the paper addresses the UHI phenomenon to adapt to urban climate change, reduce carbon emission, increase thermal comfort and health of inhabitants, and re-establishing high-quality urban life.

## The UHI Effect and Vulnerability in Compact Urban Areas

The unplanned areas have been associated with the rapid growth of urban areas in cities. This is a significant problem in the overpopulated cities in developing countries like Egypt. The unplanned regions refer to the high demand for residential areas and lack of planned housing fit the individual's income (CSD, 2013). These unplanned areas are not designed within planning policies causing socio-environmental problems. One of the serious environmental problems in the high-density urban areas is raising the temperature of the built environment (Bek et al., 2018). This causes the emergence of the UHI phenomenon. The thermal balance of the urban areas is one of the factors of UHI occurrence due to several reasons such as lack of green/cooling public spaces and conversion of greenery into impervious surfaces. Many studies (Shahmohamadi (Kloss) et al., 2011; Gratland, 2008; Oke, 1982) have demonstrated that higher temperatures develop in areas that contain a high percentage of low albedo, water-resistant surfaces as well as a low percentage of greenery and moisture trapping characters. Based on urban characteristics of cities, the most considerable elevations of urban temperature occur during clear nights, which is usually between 3 and 5°C and in some cases can also reach between 8 and 10°C (Givoni, 1998). Figure 1 shows a UHI profile. Measurements also show that the surface UHI can be as high as 15°C difference (Santamouris, 2001).

The negative consequences of the UHI increase water demand (Guhathakurta and Gober, 2007), mortality and morbidity risk due to heat stress (Tan et al., 2010), energy consumption through cooling requirements (Hirano and Fujita, 2012), reducing air quality (Plocoste et al., 2014), and have a strong negative influence on natural ventilation potential (Ghiaus et al., 2006). Moreover, there is a direct relationship between infection and air temperature. In Osaka, Japan, it has been proven that the

*Figure 1. UHI profile*
*Source: EPA, 2008*

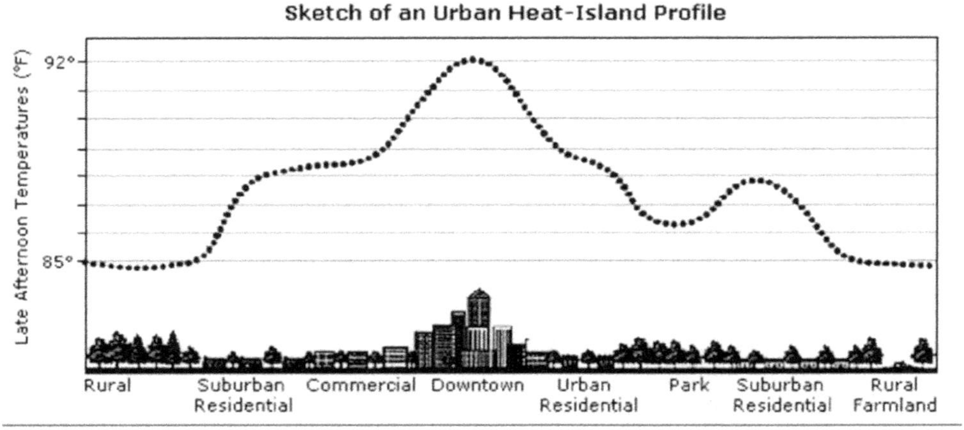


number of daily reports of disease increased when the average outdoor temperature exceeded 22°C (Narumi, 2007).

The potential drawbacks of UHI require the development of mitigation strategies. Several studies have developed the mitigation strategies, such as the modification of a building and surface material (Tsoka, 2017), the alteration of urban morphology (Algeciras et al., 2018), the installation of irrigation systems (Broadbent et al., 2018), and the inclusion of green infrastructure in the planning of cities (Wu and Chen, 2017). In high-density urban areas, the interconnected green network, including the natural and designed greeneries-from forests, agricultural lands, parks, street trees, urban gardens, green walls, and roofs to privately owned gardens, contributes to modifying the thermal balance of the urban areas and benefits the inhabitants' mental and physical well-being through cooling and providing walking opportunities.

Greenery measures mitigate the adverse effects of the UHI through shading and evapotranspiration. Shading blocks solar radiation and prevents the raising of land surface temperature. Evapotranspiration refers to absorbing the solar heat and converting it into the latent heat of evaporation, which contributes to the cooling of the surrounding areas. Greeneries significantly benefit the cooling air temperature in the arid and semiarid urban areas (Armson et al., 2013). These greeneries, such as the network of oases, benefit the city through the energy modification in the heated urban area. Canopy cover defines the temperature in parks, streets, around the buildings and the health status of the greeneries is essential in the efficient cooling of the urban areas (Chellamani et al., 2014). However, the benefits of the green spaces to microclimate vary by size, location, and types of vegetation (Long et al., 2018).

The U.S. Environmental Protection Agency (2007) stated that standard black asphalt roofs, metal or metallic roofs, and cool roofs could reach 74–85°C, 66–77°C and 43–46°C respectively at midday during the summer. Conventional roofs can be 31–47°C hotter than the air on any given day, while cool roofs tend to stay within 6–11°C of the background temperature.

Some studies have highlighted the benefits of green roofs and walls, such as reducing the temperature, which in turn reduce the cooling demand of buildings in the summer (Razzaghmanesh et al., 2016) and the need for winter heating (Takebayashi and Moriyama, 2007), contributing to air quality improvement (Baik et al., 2012; Yang et al., 2008), enhancing the energy performance of buildings (Perini and Magliocco, 2014), and reducing stormwater runoff (Jim, 2015).

A few studies have attempted to measure the UHI effect in cities in Egypt (El-Magd et al., 2016; Shahraiyni et al., 2016; Abutaleb et al., 2015; Burri et al., 2009; Ghoneim, 2009; Robaa, 2003). There is much less effort dedicated to assessing the urban climate in Cairo with regards to the different areas and related morphologies and landscapes. An urban climate map was developed for Cairo to identify the different climates within the urban agglomeration (Khalil et al., 2018). The map draws on an analysis of land use, urban structure, vegetation, roughness (change in heights). Based on experts' knowledge and available data, various local climatopes can be identified (Ren et al., 2011). It is no surprise that most of Cairo's dense neighbourhoods suffer from increased heat stress and reduced ability to self-heal. Among these highly dense areas are unplanned areas with compact buildings and a lack of open spaces. Thus, reducing the potentialities of ventilation and increasing the heat storage within the buildings, the latter accentuated by the utilization of low albedo materials (Khalil et al., 2018). Although a climate change strategy has been developed for Giza (part of Greater Cairo), it hasn't studied UHI in the city or its urban climate (Giza Governorate, 2018). The required institutional capacities are lacking and needed resources, which have resulted in the lack of implementation of almost any of the proposed strategy projects (Eissa & Khalil, 2021).

# METHODOLOGY

The methodology is mainly based on secondary data collection, observation, and remote sensing to analyse the problem in the target area.

## Secondary Data Collection and Observation

Secondary data collection for this research is based on a literature review. The literature review formed part of an essential process of conducting the study. It serves to propagate the formulation of the research problems and identify the boundary of knowledge. Moreover, observation as the data collection method was used to complement the preliminary study.

## Remote Sensing

Google Earth Engine Open-Source Code for LST estimation from the Landsat series-Remote Sensing (Ermida et al., 2020) has been applied to map LST. Since the hottest days of Cairo are in the first two weeks of July, the LST has been mapped for the date range of 1st-15th July 2020. The satellite image was derived from Landsat 8.

In addition, NDVI and NDBI have been also mapped to identify the existing greeneries and the relation between hotspots and land coverage and the potential vacant lands for intervention.

The 1C-level-processed, cloudless Sentinel-2A image of July 2020 was used to estimate the NDBI and NDVI, for the Greater Cairo Region. The sentinel-2 image includes 13 spectral bands: four visible and near-infra-red bands at 10 m, six bands at 20 m and three bands at 60 m spatial resolution. NDBI has been mapped using near-infra-red (Band 8) and red (Band 4) 10 m spectral bands. The NDVI equation is as follows:

$$NDVI = (NIR(B8) - RED(B4)) / (NIR(B8) + RED(B4))$$

This index outputs values between -1.0 and 1.0, primarily representing greenness. Moderate values (0.1 to 0.5) represent shrub and grassland, while high values (0.5 to 0.8) indicate temperate and tropical rainforests. Any negative values are mainly generated from clouds, water, and snow, and values near zero are generated mainly from rock and bare soil. Shallow values (0.1 and below) of NDVI correspond to barren areas of stone, sand, or snow.

The NDBI highlights urban areas with higher reflectance in the shortwave infrared spectral range (SWIR). NDBI has been mapped using SWIR (Band 11) and red (Band 8) spectral bands of the Sentinel-2A image. SWIR, 20 m resolution band of the Sentinel-2A imagery was resampled to 10 m by using the nearest neighbour method to maintain the spatial resolution integrity. The NDBI equation is as follows:

$$NDBI = (SWIR(B11) - NIR(B8)) / (SWIR(B11) + NIR(B8))$$

NDBI values range from -1 to 1. -1 to 0 represents water and green areas, -0.1 to 0.1 represent barren areas, and 0.1 to 1 represent built-up areas. Google Earth was used for visual interpretation and validation of the results.

# DISCUSSION

## The UHI Effect in Unplanned Areas of Greater Cairo

Rapid Urbanization in Cairo multiplied unplanned settlements. Almost 70% of Greater Cairo is random (Sims, 2012; Howeidy et al., 2009). Generally, these areas are struggling with inadequate infrastructure facilities and essential services. In addition, plenty of the unplanned regions has been built highly compact with paying very little attention to the extreme local climatic conditions. On the other hand, high dense built-up areas crowd out greenery and sealed urban areas with low albedo materials, which increase absorption of solar radiation through urban surfaces. As a result, the city notably suffers from UHI effects, higher energy consumption, and over-generation of greenhouse gas emissions. These changes have the potential to have severe consequences for the health and well-being of inhabitants, especially in unplanned areas where they are more vulnerable to climate change and have a lack of access to appropriate facilities and services. Living in poverty and alone, being physically or mentally ill, etc., makes unplanned areas more vulnerable. People in poverty cannot escape the heat due to a lack of financial resources to make necessary changes, such as purchasing air conditioners or making appropriate home renovations.

UHI definition can differ in different scales, such as regional, city, and neighbourhood. On a regional scale, sealed urban areas of a metropolitan region constitute a UHI in their entirety. At the scale of a city, certain areas include UHIs compared with their surroundings. At the neighbourhood scale, low albedo material of roads or roofs of buildings may constitute a UHI compared to a piece of land covered with greenery (Figure 2). There are many intervening scales of analysis, but in this paper, the UHI effect is analysed at the city and neighbourhood scales to identify the hotspots that are at particularly high risk to UHI.

## Factors Contributing to UHI Intensification in Cairo

To develop effective UHI adaptation strategies, it is first necessary to determine what factors play a crucial role in intensifying UHI in Cairo and cause inhabitants to be exposed and vulnerable to heat. Based on the observation and literature analysis, there are plenty of factors that can contribute to intensifying the UHI effect. In the case of Cairo, the factors can be divided into two categories:

### Urban Structure/Physical Factors

Many features of the physical structure of a city affect its urban climate and cause microclimate changes such as:

*Figure 2. Geographical location of the area with the highest heat hazard in three scales: regional, city and neighbourhood*
*Source: Authors*

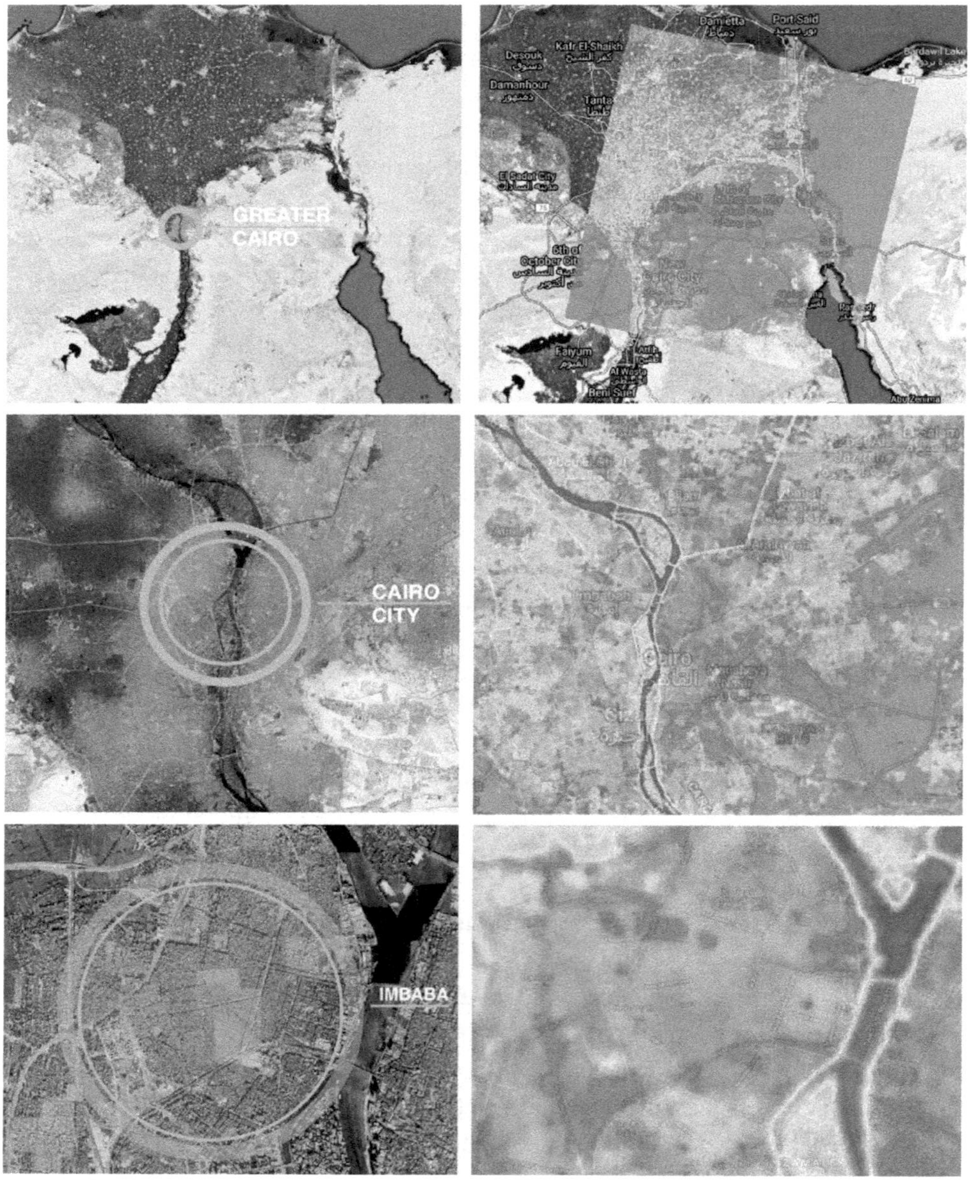


## 1. City Location

Greater Cairo is surrounded by desert from the South, West and East (Figure 1), hence considered the biggest arid city. The deserts are entirely exposed to direct sun, which increases the absorption of solar radiation and creates high surface temperature. Re-emission of the solar radiation to the ambient air at night intensifies ''Air UHI'' in the entire Greater Cairo and forms a dome of stagnant warm air over the region. It is as if the city is located in the inner core. Moreover, it has many other critical/ hotspots by itself, and it is somehow struggling with heat stress from all sides, which makes the current situation acute.

## 2. Size of the City and Population

The boundaries of the built-up area of Greater Cairo is 3133 Km2 (GOPP, 2012), and the total population estimates more than 21 million with an annual growth rate of 2.02% (Macrotrends, 2021). The population density varies within districts of Greater Cairo but is mainly highest in the city core and its informal areas (Figure 3).

City and population size can substantially affect the magnitude of the UHI. Rapid urban expansion creates sealed urban areas on a large scale, increasing solar radiation absorption. Moreover, in large cities like Greater Cairo, driving private cars increases fuel consumption and generates more anthropogenic heat contributing to the UHI effects.

## 3. The Density of the Built-Up Area

The built-up ratio is high primarily within the city's urban core and specifically in unplanned areas, reaching 85%. Figure 4 shows the distribution of the built-up ratios around Greater Cairo.

High urban density, especially in unplanned areas, can change the microclimate through:

1. Land covered by buildings modifies wind, radiant balance, and temperature conditions near the ground level.
2. Distance between buildings within an urban block significantly affects ventilation conditions, both outdoors and indoors; and
3. Higher buildings reduce ground-level wind speed more than lower buildings.

## 4. Urban Geometry

In very tight unplanned areas, urban canyons partly block the skydome's view, which restricts the cosmic cooling effect at night and reduces air ventilation while simultaneously multiplying the reflection of long-wave solar radiation (Figure 5). The trapped solar radiations in urban canyons change the microclimate contributing to the UHI effect.

*Figure 3. The population density varies from 50 to 2500 ca/feddan (200 to 10000 ca/ha)*
*Source: GOPP, 2014*

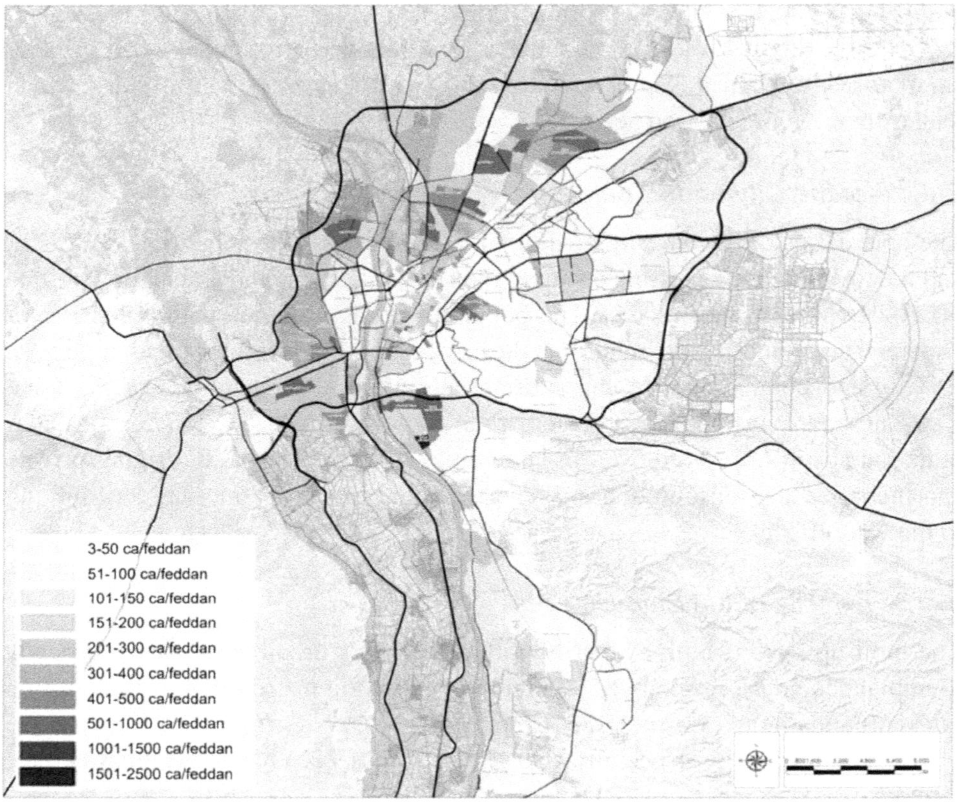


## 5. Thermal Properties of the Fabric

Concrete, brick, and asphalt are the predominant construction materials in Cairo that have higher heat storage capacity and thermal inertia and absorb and retain more solar radiation than natural materials. It is vital to note that buildings in unplanned areas are of low thermal insulation, especially in the case of makeshift materials. In addition, concrete and brick buildings, which comprise the majority of these areas, usually lack any plastering or painting, thus rendering the indoor temperature very high due to the high U value. Furthermore, the low albedo materials, dark-coloured bricks, increase heat stress in microclimate and eventually increase the UHI effect (Figure 6).

*Figure 4. The built-up ratio in different parts of Greater Cairo*
*Source: Authors, based on GOPP, 2014*

### 6. Surface Waterproofing

In Cairo's compact, unplanned urban area, greenery is scarce, although it can significantly contribute to improving the microclimate. Lack of porosity materials and vegetated and moisture trapping surfaces create an evaporation deficit in the city. Planting a tree remains an individual initiative and subject to one's ability and commitment to maintain the tree and ensure its prosperity.

### 7. Building Typology and Condition

The flat roofs in Cairo and low albedo materials like asphalt, can trap the heat by being more exposed to heat, increase land surface temperature and worsen health risks for inhabitants (Figure 6). The physical condition of buildings is also crucial in increasing the chance of being exposed to heat. Mainly, people living in unplanned areas and on top floors apartments are more at risk due to lack of services, inappropriate building conditions constructed without paying attention to climatic factors and poor insulation.

*Figure 5. The complex urban geometry of Cairo City*
*Source: Authors*

## Non-Physical Factors

### *1. Anthropogenic Heat*

Human and economic activities in the city, such as manufacturing, markets, traffic congestion, and lighting, are responsible for a high percentage of anthropogenic heat release (Figure 7). The heat from these sources warms the ambient air by conduction, convection, and radiation, increasing UHI effects.

### *2. Air Pollution*

UHI effects increase the demand for air-conditioning, causing higher local and regional air pollution through power generation from fossil fuels. Increased traffic congestion in Cairo contributes to a high rate of air pollution (Figure 8). Cairo's air pollution has already reached unhealthy and dangerous levels, mainly affecting children and the elderly.

*Figure 6. Building typology in the unplanned urban areas*
*Source: Authors*

*Figure 7. Food markets in Cairo*
*Source: Authors*

### 3. Lack of Awareness and Inappropriate Lifestyle

Lack of awareness and the inappropriate lifestyle of most inhabitants also multiply the described challenges. A resource-intensive consumer society driven urban lifestyle contributes significantly to the pressure on the ecosystem (Figure 9). Capacity building and raising awareness should be significantly considered in parallel with physical intervention.

### 4. Meteorological Factors

According to the World Bank, Egypt is suffering from an increase in tropical nights (with a temperature of more than 20°C) and is expected to continue under various emission scenarios. The threat of increased tropical nights is closely associated with health risks as organisms can generally withstand high temperatures during the day only, given enough cooling takes place at night (World Bank Group, 2021). This is even accentuated with the UHI effect within compact cities such as Cairo.

*Figure 8. Traffic congestion in Cairo*
*Source: Authors*

## Case Study Area: Imbaba

Imbaba is selected as it is a prototype for highly dense unplanned areas in Greater Cairo. To a great extent, what works in Imbaba would work in many other similar situations, even in other Egyptian cities along the Nile and Delta. Additionally, we had access to data and familiarity with the socio-economic and environmental conditions through previous projects, which addressed our time and budget limitations.

Cairo shows a frequent dominant northerly component, where wind speed is sufficient to penetrate the Urban Canopy Layer (Robaa, 2003). Fortunately, this direction is congruent with the path of the river Nile northwest allowing for ventilation paths and other north-easterly air paths facilitated by existing corridors. The Khamasin (South-westerly wind with dust) presents a challenge to the city requiring shelter. During a sandstorm, the dust in the air can change the urban net all-wave radiation budget and multiply the reflection of long-wave solar radiation. Figure 10 shows the climate analysis for Imbaba.

*Figure 9. Urban waste has been left in an unplanned area of Cairo*
*Source: Authors*

*Figure 10. The climate analysis for Imbaba*
*Source: Authors*

Unplanned areas are either built on state-owned land or private land. The first has a very compact irregular original pattern, mostly low-rise buildings. Some of the buildings are made of makeshift materials. Generally, people are of low income in this area. The majorly prevailing typology of the unplanned regions in Imbaba is those built on former private agricultural land that was illegally subdivided. The urban pattern follows the former geometry of agricultural basins with very narrow streets and extra-long blocks. Most buildings range from 4-6 floors high with mixed uses. Most buildings are of skeleton structure and brick walls lacking any plastering or painting (Figure 11a and 11b). These areas vary in income but can have low to lower-middle-income households with some middle-income households.

Imbaba district, with one million inhabitants, extends back to the 1960s. It was originally agricultural land and was still classified as rural until the 1950s, then turned to an informal area since the first wave of informal construction in 1950–1977. The district is informally built, advocating walkability, mixed use, and mixed housing. Transportation is integrated with land use. Less than 10% of buildings are good quality, while 85% are of moderate quality. There are almost no green spaces within the area, with only 0.05 m2/ca of open spaces (Khalil and Gammaz, 2020),

*Figure 11. a) Building typology; b) Brick wall skeleton of the buildings*
*Source: Authors*

excluding the newly opened 9 ha Imbaba park. Figure 12 shows the NDVI map in Imbaba. Some greenery has just remained in the West and South parts of Imbaba. The foliage has been already crowded out in the centre of the area and covered with compact and heavily built-up areas. This meagre ratio of green spaces is a resultant of the informal process of urban growth where no margin is reserved as a response to economic pressures. However, the quality of life in Imbaba cannot be termed as acceptable (Figure 13). The lack of adequate services, education, health, youth facilities is expected. In addition, the long narrow streets decrease connectivity in the area due to incompatibility with motorized vehicles leaving the only choice to auto-rickshaws "Tuktuks" (Figure 14). The extensive reliance on low-quality microbuses and pickup trucks to transport people is another hazard that increases pollution. The scarcity of planned open spaces has led to the contestation of the existing public space where streets are used as public spaces for socializing and vending (Figure 15). Finally, the extensive use of bricks, concrete and asphalt (within wide streets) contributes to increasing UHI and increasing thermal loads on buildings (Attia and Khalil, 2015).

Although these areas are usually viewed negatively, they possess many aspects of sustainable urbanism, with compactness and mixed uses at the forefront (Khalil and Gammaz, 2020; Khalil and Khalil, 2015). Interestingly, in some areas, people have

*Figure 12. NDVI map of Imbaba*
*Source: Authors*

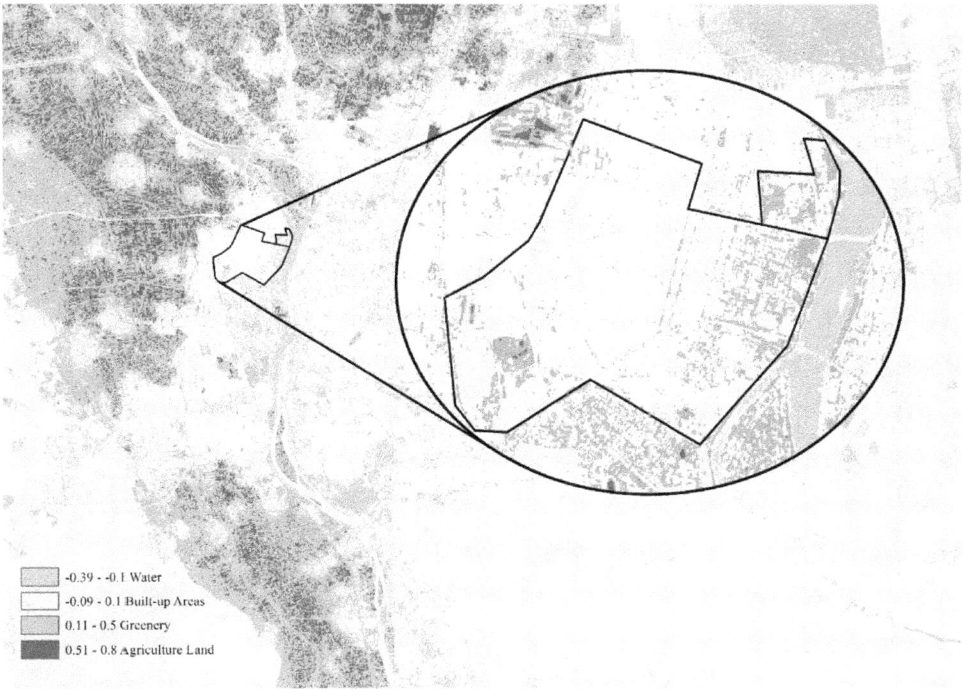


devised some simple interventions like installing shading devices in public spaces with high use, whether occasionally or permanently. The example is given here for the shading for the Friday prayers area or a specific market (Khalil et al., 2018). These copies are from the old city urban features as in the famous Khyamia Area in Islamic Cairo. Other similar interventions include shading windows widely using fenestrations and canvas awnings or rarely using traditional wooden mashrabiya.

The LST has been mapped to identify and prioritize the hotspots in Imbaba (Figure 16). The map shows that the hottest spots have been distributed in the whole Imbaba, mostly located in the West, South and North, with temperatures between 45°C and 51°C. Greeneries are the coolest spots with temperature between 31°C and 36°C because of their less heat capacity, which absorbs and retains less solar radiation.

NDBI has also been mapped to find the relation between hotspots and land coverage. By overlaying the LST and NDBI maps (Figure 17), the vacant lands and parking lots are prioritized as one of the hottest spots due to being more exposed to solar radiation. The heavily compact built-up areas in the centre of Imbaba are the second hotspots in the prioritization list due to the absence of greenery and lack of access to cooling public spaces with temperature between 40°C and 45°C. This is

*Figure 13. Quality of life in Imbaba*
*Source: Authors*

*Figure 14. Narrow streets in Imbaba*
*Source: Authors*

*Figure 15. Streets as public spaces for vending in Imbaba*
*Source: Authors*

due to being completely sealed off with low albedo and water-resistant materials, which have a high potential of absorbing solar radiation. They are in second place because the shadow of the buildings makes them less exposed to solar radiation. Hence, the areas of intervention have been chosen based on identified hot spots as priority spots to take action.

*Figure 16. LST map for Greater Cairo and Imbaba Area*
*Source: Authors*

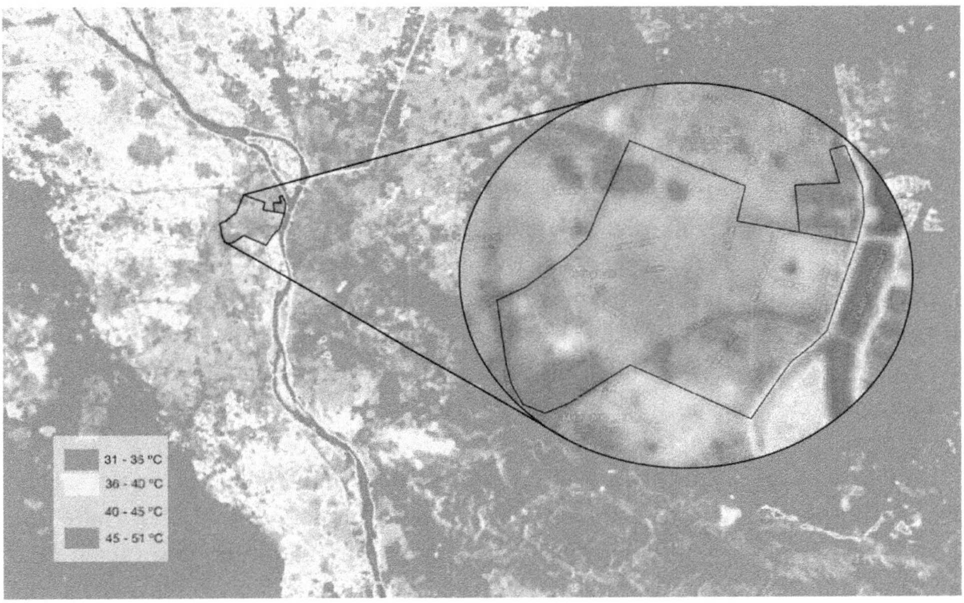


## RECOMMENDATIONS

To tackle the UHI effect in the target area, in a framework of passive urban cooling, designing a chain of cooling public spaces as breathing points in a very compact area of Imbaba are suggested. The starting point of the designed cooling public space is from a vacant land along the Nile, as one of the most important potentials of Cairo, followed by further interventions into the target area and extended into alternative vacant lands at the North and South parts of Imbaba.

This approach resembles an urban acupuncture approach (Lerner, 2014), where targeted local interventions complemented to have an overall positive effect. The use of an acupuncture approach becomes of significant benefit in hyper-dense situations as expressed in the context of Cairo, specifically its informal/unplanned areas where

*Figure 17. Overlaying NDBI (a) and LST (b) maps presenting the relation between hotspots and land coverage*
*Source: Authors*

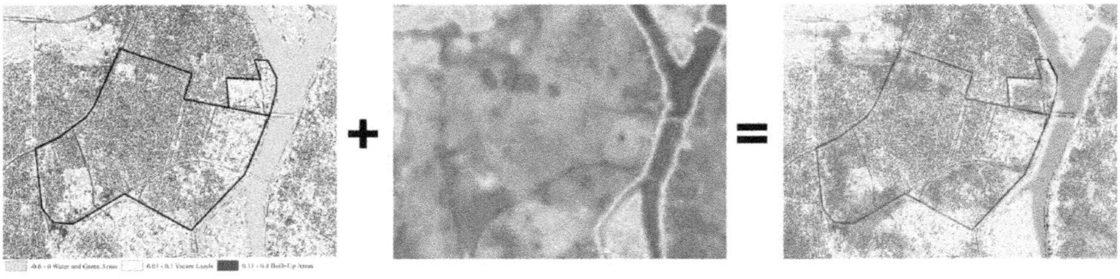

densities could reach up to 1800 persons/hectare (Khalil and Gammaz, 2020). In such localities, deploying a series of interlinked interventions would be much more suitable than looking for extensive scale interventions given the intricacy of urban structures, land ownerships and related livelihoods.

A wooden structure inspires the central concept of the proposed public space as one of the traditional architectural elements in Egypt called "*Takeeba*". Takeeba is usually covered with plants to provide shade and a breeze for cooling down and relaxing on sweltering days of summer. Traditional materials are used where palm trunks and fronds are standard (Figure 18). This concept has been used in the whole design process from details to generalities.

Since there are several types of UHI, which occur in different horizontal scales, distributing greenery onto several levels can tackle different types of UHIs. Therefore, the concept is to start with elevated greenery on one of the hottest vacant land adjacent to the Nile River by designing light and prefabricated structures with natural and high albedo materials, continued through Imbaba, where the hottest spots are located, by integrating greenery, including green corridors, small cooling public spaces, vertical and horizontal greenery, etc., eventually ending up with another elevated vegetation in another hottest vacant land near the Imbaba Park (Figure 19).

Integrating greenery into the urban area can interrupt microclimate changes in the urban canyon decrease the anthropogenic heat and the energy consumption in the surrounding buildings as well. This chain of integrated greenery increases evapotranspiration in a compact urban area of Imbaba and contributes to creating lower temperature spaces.

Moreover, as greenery covers the skydome of urban areas, the urban surfaces are less exposed to solar radiation. Thereby a temperature reduction can be expected, according to the study done by Shashua-Bar and Hoffman (2002). Furthermore, Gomez et al. (2001) declared that the bigger the urban green pocket, the more pronounced thermal comfort improvement would be. Additionally, trees planted over green areas

*Figure 18. Different setups of Takeeba*
*Source: from top left to right bottom, (Hassan & Amr, 2018; curtesy of Kosta Mathey; DIY Vine Arch, n.d.; DIY vine sheds, 2020)*

can transpire better than asphalt areas (Kjelgren and Montague, 1998). Santamouris et al. (2001) declared that unshaded and shaded asphalt temperatures were around 52°C and 35°C, respectively. At the same time, the temperature of shaded white slabs used in pavements was between 28°C and 31°C with an ambient temperature of 31°C. Jauregui (1993) found a 5.5°C air temperature reduction within a two-mile radius of Mexico City's urban parks.

The interconnected cooling public spaces act as a huge cool island in the target urban area and provide thermal comfort. Besides greenery, materials, especially their colours, play a crucial role in the formulation of the thermal environment inside urban canyons. Gartland (2008) proved that the temperatures of dark and dry surfaces like asphalt and concrete in direct sun could reach up to 88°C during the day, while greenery with the same conditions might reach only 18°C. This presents the temperature differences between urban and natural materials. Therefore, the design applied natural and high albedo materials to avoid the absorption of solar

*Figure 19. Alternative design proposals*
*Source: Authors*

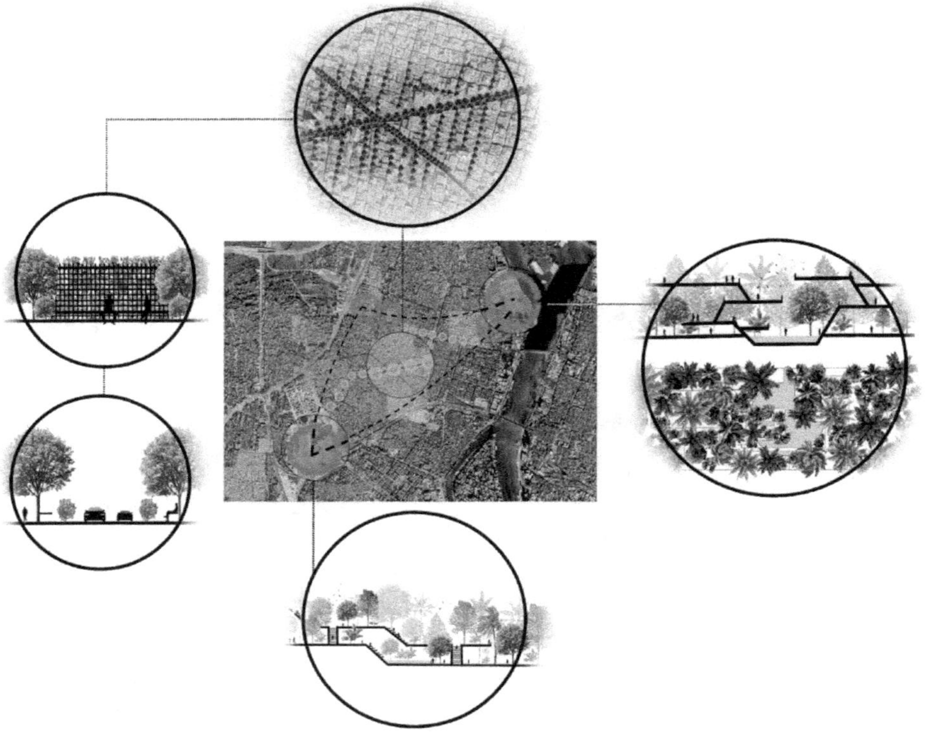

radiation. The structure of the alleviated greenery has been made by white-coloured wood to increase the reflectivity during the day and send the solar radiation to the atmosphere rather than accumulating them into the urban canyon. Therefore, the temperature can be reduced, and thermal comfort can be provided by placing the cooling public spaces on the urban canyons of the target metropolitan area. This will avoid restricting the cosmic cooling effect at night, reducing air ventilation, and multiplication of the reflection of long-wave solar radiation.

The existing urban canyons partially block the sky dome's view and block the air's free movement. The reradiated solar radiation can be trapped into the urban canyons and change the micro-climate through the multiplication of solar radiation. Lower wind speeds and higher temperature within urban canyons mean a local accumulation of heat in buildings, resulting in increased energy consumption and CO2 production– which on their turn into 'urban radiators', heat the ambient air and add more pollution into the urban canyons.

Besides all these, water and irrigation systems are also essential to keep the cooling public spaces always green and alive. Several ways can be employed to irrigate the greenery. The main one is ''using greywater generated by surrounding buildings''. Connecting the irrigation system with the existing greywater collector system of the surrounding building can solve the water shortage. Sources of greywater are sinks, showers, baths, washing machines or dishwashers. As greywater contains fewer pathogens than domestic wastewater, it is safer to handle and easier to treat and reuse onsite for landscape or crop irrigation and other non-potable uses. Tailoring modern architecture and state of the art technology to the local context is also considered.

The particles in the air can change the urban radiation budget by re-emitting long-wave radiation from the urban surface back to the ground and reabsorb by the urban surfaces and increasing the ambient air. The pollution contributes to the intensification of the UHI effect. In addition, using a mixture of local trees, bushes, smaller plants, and flowers can also be extremely helpful in reducing air pollution and improving air quality.

Trees absorb gaseous pollutants through leaf stomata and can dissolve or bind water-soluble pollutants onto moist leaf surfaces, while tree canopies intercept particulates. In parallel, trees reduce ozone concentrations in the ambient air, either by directly absorbing the ozone (and other pollutants like NO2) or by reducing air temperatures, reducing hydrocarbon emission and ozone formation rates (McPherson et al., 1998). Bernatsky (1978) reported that a street lined with healthy trees can reduce airborne dust particles by around 7000 particles per litre of air. Trees also remove atmospheric CO2 and store it as a woody biomass

Providing high-quality open spaces and urban thermal comfort for the citizens can also protect them from the UHI effect, especially vulnerable people. The chain of cooling public spaces revitalizes the streets for walking, gathering and recreation. Skyrise greenery can provide enjoyment with the landscape defining the area and providing inhabitants with the setting for outdoor recreational pursuits, especially the elderly, handicapped and the young who are more dependent on near-home recreation spaces. Through skyrise greenery, cities can be made more liveable by providing significant amounts of accessible outdoor recreation or amenity space. Given the high premium for street-level areas in the towns, elevated greenery offers opportunities for more secluded, less polluted, and less noisy spaces for informal recreation.

Utilizing the different greenery levels and its gardens as terraces for human interaction can contribute to the renewal of an urban area. The structures will provide maximum shading, good viewpoints, and cosy places to relax and interact. Visual and physical contact with plants can result in direct health benefits. Plants can generate restorative effects, leading to decreased stress, improving patient recovery rates, and providing higher illness resistance (Ulrich and Persons, 1992).

The effect of greenery on people from different points of view can be summarized in the following way:

1. Residents with balcony or terrace gardens are less susceptible to illness in a high-density environment (Johnston and Newton, 1996)
2. The World Health Organization (WHO, 2012) recommended the availability of a minimum of 9 m2 of green space per individual with an ideal green space value of 50 m2 per capita
3. The 'sensory dimensions' of greenery can also lead to the health improvement of its users (Dwyer et al., 1994)
4. High temperatures can have numerous adverse effects on the physical and psychological well-being of the population. Unnaturally high heat loads can, directly and indirectly, reduce life expectancy. The potential of skyrise greenery to lower temperatures can reduce heat-aggravated illness and death among the city population (Weihe, 1986)

Therefore, according to the above perspectives, there is no doubt that plants play a significant role in providing thermal comfort within cities for their inhabitants. It lets them enjoy a comfortable, cool zone and increase the inhabitants' satisfaction and their health. Besides, the cooling public spaces offer a real surprise in the urban setting for people who stumble across it accidentally and become a meeting point for various people. These surprises can act as a magnet and distract people from the harsh sealed urban areas with hot climates.

It is an urban landmark that surprises pedestrians and attracts tourists through its unusual and utopian appearance. One of our concerns was to create a public space that improves the aesthetics of the urban area. Landscaping has often been used to enhance the aesthetics of the urban area. The support for the preservation of plants has been attributed to the attraction that many urbanites feel for a natural landscape. They also add value to Cairo and the neighbourhood in which it is located, providing identity and a site for cultural activities.

On the other hand, the proposed cooling public spaces can also have an economic benefit, which will be the decrease of energy consumption of the surrounding buildings as they have a cool down effect on its close neighbourhood. Gaudet (1985) declared that each degree (°C) of heat requires an additional 3-4% of cooling energy. He also proved that 6°C reduction in air temperature through the adoption of shade trees might reduce the consumption of air-conditioning by 50-70% in surrounding buildings.

Another substantial economic benefit is the rising land value around parks or greenery, as people usually look for apartments near vegetation areas. Also, the pleasant environment provided by the proposed cooling public spaces invites the locals to spend their time around the space, which opens up chances for shops,

*Figure 20. Future plan for tackling UHI in Cairo*
*Source: Authors*

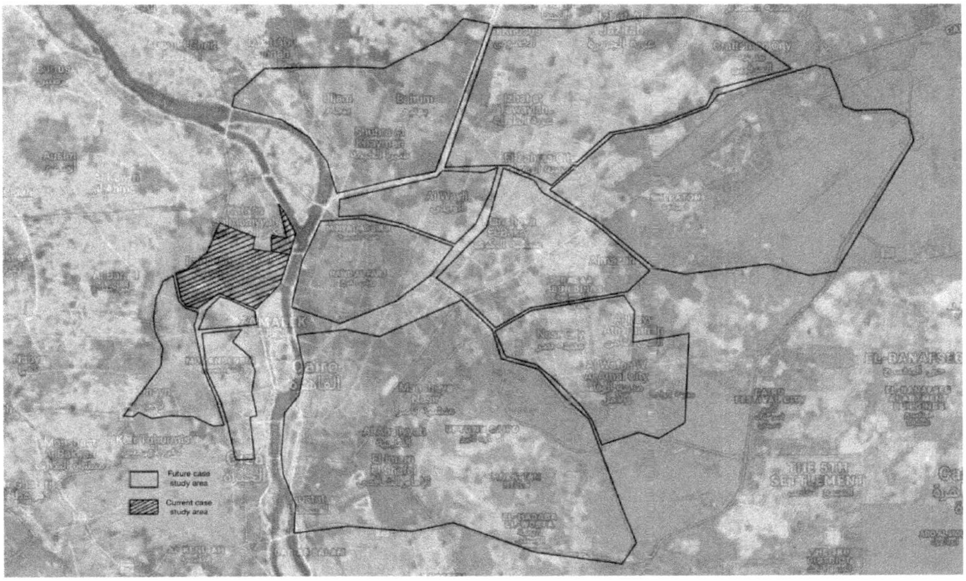


cafes, and restaurants to increase their businesses nearby. Additionally, open-air markets, music events or sports groups can benefit from the cooler temperatures and shading and increased pedestrian traffic. For instance, they can organize events and festivals in the urban area during the weekend and at the same time enjoy the pleasant environment.

## THE WAY FORWARD

It is impossible to tackle the UHI effect in the entire Greater Cairo just by implementing a chain of cooling public spaces in an urban area. For this reason, the replicability of the idea was considered, and we decided to replicate the concept for Cairo (in both planned and unplanned areas) in a follow-up project. As seen in Figure 20, there are a lot of hot spots in Cairo that should be tackled. This can provide plenty of urban lungs to let the city breathe, cool down, and move towards urban climate change resilience, which is one of the main goals of SDGs and the NUA. Integrating with the recently developed Giza climate change strategy would be a starting point to realising UHI reduction in the city.

## REFERENCES

Abutaleb, K., Ngie, A., Darwish, A., Ahmed, M., Arafat, S., & Ahmed, F. (2015). Assessment of urban heat island using remotely sensed imagery over Greater Cairo, Egypt. *Advances in Remote Sensing*, *04*(01), 35–47. doi:10.4236/ars.2015.41004

Algeciras, J. A. R., Tablada, A., & Matzarakis, A. (2018). Effect of asymmetrical street canyons on pedestrian thermal comfort in the warm-humid climate of Cuba. *Theoretical and Applied Climatology*, *133*(3-4), 663–679. doi:10.100700704-017-2204-8

Armson, D., Rahman, M. A., & Ennos, A. R. A. (2013). Comparison of the shading effectiveness of five different street tree species in Manchester. *Arboriculture & Urban Forestry*, *39*(4), 157–164. doi:10.48044/jauf.2013.021

Attia, S., & Khalil, H. (2015). Urban Metabolism and Quality of Life in Informal Areas. In *Proceedings REAL CORP 2015: Plan Together – Right Now – Overall*, 661-674.

Baik, J., Kwak, K. H., Park, S. B., & Ryu, Y. H. (2012). Effects of building roof greening on air quality in street canyons. *Atmospheric Environment*, *61*, 48–55. doi:10.1016/j.atmosenv.2012.06.076

Bek, M. A., Azmy, N., & Elkafrawy, S. (2018). The effect of unplanned growth of urban areas on heat island phenomena. *Ain Shams Engineering Journal*, *9*(4), 3169–3177. doi:10.1016/j.asej.2017.11.001

Bernatsky, A. (1978). *Tree Ecology and Preservation*. Elsevier Scientific Publishers.

Broadbent, A. M., Coutts, A., Tapper, N. J., & Demuzere, M. (2018). The cooling effect of irrigation on urban microclimate during heatwave conditions. *Urban Climate*, *23*, 309–329. doi:10.1016/j.uclim.2017.05.002

Burri, S., Frey, C., Parlow, E., & Vogt, R. (2009). CO2 fluxes and concentrations over an urban surface in Cairo/Egypt. *In Proceedings of The Seventh International Conference on Urban Climate*. Yokohama, Japan.

Center for Sustainable Development (CSD). (2013). *Egypt's Strategy for Dealing with Slums*. American University in Cairo.

Chellamani, P., Singh, C. P., & Panigrahy, S. (2014). Assessment of the health status of Indian mangrove ecosystems using multi temporal remote sensing data. *Tropical Ecology*, *55*, 245–253.

DIY vine arch. (n.d.). Retrieved February 25, 2022, from https://www.cerceis.com/1432/

DIY vine sheds. (2020, February 5). *DIY garden*. https://ar.imadeself.com/ yagody/karkas-dlya-vinograda.html

Dwyer, J. F., Schroeder, H. W., & Gobster, P. H. (1994). The Deep Significance of Urban Trees and Forests. In R. H. Platt, R. A. Rowntree, & P. C. Muick (Eds.), *The Ecological City: Preserving and Restoring Urban Biodiversity*. University of Massachusetts Press.

Eissa, Y., & Khalil, H. A. (2021). Urban Climate Change Governance within Centralised Governments: A Case Study of Giza, Egypt. *Urban Forum*. doi:10.1007/s12132-021-09441-9

El-magd, I.A., Ismail, A. & Zanaty, N. (2016). Spatial variability of urban heat islands in Cairo City, Egypt using time series of Landsat Satellite images. *International Journal of Advanced Remote Sensing and GIS*, *5*, 1618.

Ermida, S. L., Soares, P., Mantas, V., Göttsche, F. M., & Trigo, I. F. (2020). Google Earth Engine open-source code for Land Surface Temperature estimation from the Landsat series. *Remote Sensing*, *12*(9), 1471.

Gartland, L. (2008). *Heat Islands: Understanding and Mitigating Heat in Urban Areas*. London Earthscan publishing.

Gaudet, C. (1985). Sunspots, landscaping for energy efficiency. *Harrowsmith*, *1*(61), 20–33.

General Organization of Physical Planning GOPP. (2009). *Cairo Future Vision 2050*. Author.

General Organization of Physical Planning GOPP. (2012). *Greater Cairo Urban Development Strategy*. Author.

*General Organization of Physical Planning GOPP*. (2014). Ministry of Housing, Utilities and Urban Communities.

Ghiaus, C., Allard, F., Santamouris, M., Georgakis, C., & Nicol, F. (2006). Urban environment influence on natural ventilation potential. *Building and Environment*, *41*(4), 395–406.

Ghoneim, E. (2009). A remote sensing study of some impacts of global warming on the Arab region. In M. K. Tolba & N. Saab (Eds.), *Arab Environment: Impact of Climate Change on the Arab Countries* (pp. 31–46). Arab Forum for Environment and Development.

Givoni, B. (1998). *Climate Considerations in Building and Urban Design*. John Wiley & Sons.

Gomez, F., Tamarit, N., & Jaboloyes, J. (2001). Green zones, bio-climatic studies and human comfort in the future development of urban planning. *Landscape and Urban Planning*, *55*, 151–161.

Governorate, G. (2018). *Towards a Climate Change Strategy in Giza Governorate Framework Document*. Giza Governorate. Retrieved from https://egwuf.files.wordpress.com/2020/01/giza-climate-change-strategy.pdf

Guhathakurta, S., & Gober, P. (2007). The impact of the phoenix urban heat island on residential water use. *Journal of the American Planning Association*, *73*, 317–329.

Hassan, M., & Amr, N. (2018). *Title: Palm Trees Festival First Round*. https://www.google.com/ url?sa =i&url =https% 3A% 2F% 2Fb mitp global network. org%2 Fwp- content% 2F uploads %2F 2021% 2F04% 2Fpalm- trees- festival- 2018. pdf &psig = AOvVaw 3fOOr0 FmEwST_ iEUA XgYo8 &ust = 164 590 722 329 000 0&source = images &cd = vfe & ved = 2ah UKE wjA gaW s2J v2A hVB s6Q KHW 3uC QIQ 3Yk Beg QIA BAL

Hirano, Y., & Fujita, T. (2012). Evaluation of the impact of the urban heat island on residential and commercial energy consumption in Tokyo. *Energy*, *37*, 371–383.

Howeidy, A. (2009). *Cairo's Informal Areas: Between Urban Challenges and Hidden Potentials*. GIZ Project.

Jauregui, E. (1993). *Bibliography of urban climate in tropical/sub-tropical areas*. Geneva WMO Publisher.

Jim, C. Y. (2015). Thermal performance of climber green walls: Effects of solar irradiance and orientation. *Applied Energy*, *154*, 631–643.

Johnston, J., & Newton, J. (1996). *Building Green, A Guide for Using Plants on Roofs, Walls and Pavements*. The London Ecology Unit.

Khalil, H., Ibrahim, A., ElGendy, N., & Makhlouf, N. (2018). Could/should improving the urban climate in informal areas of fast-growing cities be an integral part of upgrading processes? case study, Cairo. *Urban Climate*, *24*, 63–79.

Khalil, H., & Khalil, E. E. (2015). *Energy Efficiency in The Urban Environment*. CRC press, Taylor & Francis.

Khalil, H. A., & Gammaz, S. (2020). Supporting informal areas resilience: reinforcing hidden green potential for a better quality of life. In J. Moore, S. Attia, A. Abdel-Kader, & A. Narasimhan (Eds.), Ecocities Now: Building the Bridge to Socially Just and Ecologically Sustainable Cities (pp. 11-40). Springer.

Kjelgren, R., & Montague, T. (1998). Urban tree transpiration over turf and asphalt surfaces. *Atmospheric Environment*, *32*(1), 35–41.

Lerner, J. (2014). *Urban Acupuncture*. Island Press.

Long, X., Bei, N., Wu, J., Li, X., Tian, F., Li, X., Zhao, S., Cao, J., Tie, X., An, Z., & … . (2018). Does afforestation deteriorate haze pollution in Beijing–Tianjin–Hebei (BTH), China? *Atmospheric Chemistry and Physics*, *18*, 10869.

Macrotrends. (2021). *Cairo, Egypt Metro Area Population 1950-2021*. www.macrotrends.net

McPherson, E. G., Scott, K. I., & Simpson, J. R. (1998). Estimating cost effectiveness of residential yard trees for improving air quality in Sacramento, California, using existing models. *Atmospheric Environment*, *32*(1), 75–84.

Narumi, D. (2007). Effects of temperature change upon the regional electric power consumption. *Energy and Resources*, *28*(3), 13–19.

Oke, T. R. (1982). The energetic basis of the urban heat island. *Quarterly Journal of the Royal Meteorological Society*, *108*(455), 1–24.

Perini, K., & Magliocco, A. (2014). Effects of vegetation, urban density, building height, and atmospheric conditions on local temperatures and thermal comfort. *Urban Forestry & Urban Greening*, *13*, 495–506.

Plocoste, T., Jacoby-Koaly, S., Molinié, J., & Petit, R. H. (2014). Evidence of the effect of an urban heat island on air quality near a landfill. *Urban Climate*, *10*, 745–757.

Razzaghmanesh, M., Beecham, S., & Salemi, T. (2016). The role of green roofs in mitigating Urban Heat Island effects in the metropolitan area of Adelaide, South Australia. *Urban Forestry & Urban Greening*, *15*, 89–102.

Ren, C., Ng, E. Y., & Katzschner, L. (2011). Urban climatic map studies: A review. *International Journal of Climatology*, *31*, 2213–2233.

Robaa, S. M. (2003). Urban-Suburban/Rural Differences over Greater Cairo. *Atmosfera*, *16*(3), 157–171.

Rui, L., Buccolieri, R., Gao, Z., Gatto, E., & Ding, W. (2019). Study of the effect of green quantity and structure on thermal comfort and air quality in an urban-like residential district by ENVI-met modelling. *Building Simulation*, *12*, 183–194.

Santamouris, M. (2001). *Energy and Climate in the Urban Built Environment.* London James & James Publication.

Shahmohamadi (Kloss), P., Che-Ani, A. I., Maulud, K. N. A., Tawil, N. M., & Abdullah, N. A. G. (2011). The Impact of Anthropogenic Heat on Formation of Urban Heat Island and Energy Consumption Balance. *Urban Studies Research.*

Shahraiyni, H. T., Sodoudi, S., El-Zafarany, A., & Seoud, T. A. (2016). A Comprehensive Statistical Study on Daytime Surface Urban Heat Island during Summer in Urban Areas, Case Study: Cairo and Its New Towns. *Remote Sensing*, *8*(643). Advance online publication. doi:10.3390/rs8080643

Shashua-Bar, L. & Hoffman, M.E. (2000). Vegetation as a Climatic Component in the Design of an Urban Street. *Journal of Energy and Buildings,* 31.

Sims, D. (2012). *Understanding Cairo: The Logic of a City Out of Control.* The American University in Cairo Press.

Solecki, W. D., Rosenzweig, C., Parshall, L., Pope, G., Clark, M., & Cox, J. (2005). Mitigation of the heat island effect in urban New Jersey. *Global Environmental Change Part B: Environmental Hazards*, *6*(1), 39–49.

Takebayashi, H., & Moriyama, M. (2007). Surface heat budget on green roof and high reflection roof for mitigation of urban heat island. *Building and Environment*, *42*, 2971–2979.

Tan, J., Zheng, Y., Tang, X., Guo, C., Li, L., Song, G., Zhen, X., Yuan, D., Kalkstein, A. J., & Li, F. (2010). The urban heat island and its impact on heat waves and human health in Shanghai. *International Journal of Biometeorology*, *54*, 75–84.

The United States Environmental Protection Agency (EPA). (2008). https://www.epa.gov/heatislands

Tsoka, S. (2017). Investigating the relationship between urban spaces morphology and local microclimate: A study for Thessaloniki. *Procedia Environmental Sciences*, *38*, 674–681.

Ulrich, R. S., & Parsons, R. (1992). Influences of Passive Experiences with Plants on Individual Well-being and Health. In D. Relf (Ed.), *The Role of Horticulture in Human Well-being and Social Development: A National Symposium.* Timber Press Inc.

U.S. Environmental Protection Agency. (2008). *Heat Island Effect: Cool Pavement Product Information. Environmental Protection Agency.* https://www.epa.gov/heatisld/ images/ extra/ level3_pavingproducts.html

Weihe, W. H. (1986). *Life Expectancy and its Applications with Special Regard to Tropical Areas*. World Meteorological Organization.

World Bank Group. (2021). *Country: Egypt. Climate Change Knowledge Portal.* https://climateknowledgeportal.worldbank.org/country/egypt/climate-sector-health

World Health Organization. (n.d.). WHO Global Ambient Air Quality Database (2018). In *Atlas of Sustainable Development Goals 2020*. https://datatopics.worldbank.org/ sdgatlas/ goal -11- sustainable -cities -and- communities/ #footnote9

World Health Organization (WHO). (2012). *Health Indicators of Sustainable Cities in the Context of the Rio+20 UN Conference on Sustainable Development*. WHO.

Wu, Z., & Chen, L. (2017). Optimizing the spatial arrangement of trees in residential neighborhoods for better cooling effects: Integrating modeling with in-situ measurements. *Landscape and Urban Planning*, *67*, 453–472.

Yang, L., Lam, J. C., & Tsang, C. L. (2008). Energy performance of building envelopes in different climate zones in China. *Applied Energy*, *85*, 800–817.

## KEY TERMS AND DEFINITIONS

**Cooling Public Space:** A public space that provides breathing points for an urban area as well as thermal comfort for citizens.

**Greenery Measure:** A measure that integrates greenery into an urban area to interrupt microclimate changes in the urban canyon, decrease the anthropogenic heat and the energy consumption in the surrounding buildings as well as increase evapotranspiration and contribute to creating lower temperature spaces.

**Takeeba:** A wooden light structure covered with plants to provide shade and breeze for cooling down and relaxing in hot days of summer.

**Unplanned Area:** An area where either houses have been constructed on land that the occupants have no legal claim to or occupy illegally, or own the land but have no permit to build. The area is determined according to the strategic urban plan for each settlement according to the Egyptian Unified Building Law.

**Urban Acupuncture:** An architectural design practice of urban regeneration with the goal of having a transformative and revitalizing impact on that area and its surroundings.

**Urban Health:** The well-being and health of people who live in the cities.

**Urban Heat Island:** An urban area that absorbs and retains significant solar radiation than rural areas.

# Chapter 7
# Urban Heat Islands in Degenerative and Regenerative Cities:
## Heliopolis, Cairo, Egypt

**Hamdy Hefnawy**
*Ain Shams University, Egypt*

**Mohamed El Fayoumi**
*Ain Shams University, Egypt*

**Abeer Elshater**
https://orcid.org/0000-0002-5061-6861
*Ain Shams University, Egypt*

**Natalie Marie Gulsrud**
*University of Copenhagen, Denmark*

## ABSTRACT

*This chapter examines a community developed in the early 20th century as a sustainable development paradigm. However, today saw the implementation of particular development plans in some cities in the global south that may not align with this community's long-term goals. Two Delphi rounds and an Envi-Met microclimate simulation helped researchers of the present work use Heliopolis as a case of investigation. After a Heliopolis' urban development plan competition, this approach's results show a considerable drop in the site's initial values and ecological effect. Surface and mean temperatures and carbon dioxide levels are predicted to decline because of this significant loss. Future studies can look at how local streets become urban boundaries and context segregation because of development plans that cannot regenerate urban areas.*

DOI: 10.4018/978-1-6684-2462-9.ch007

## INTRODUCTION

Urban planning and design organise and change public spaces by reflecting the socio-morphological aspects of cities (Abusaada & Elshater, 2021b; Moore, 2020). The practice of urban design is concerned with the materiality of places, mixed uses and the changes that happen to steer them toward the desired outcome from the decision-makers' perspective (Dovey & Pafka, 2017). In some cases, stakeholders in urban projects choose interventions by selecting and combining relevant tactics and solutions to modify public spaces and their socioeconomic dynamics in ways that address regenerative design (Elshater, et al., 2022; Vandenbussche, 2018). A regenerative city is constructed on an environmentally beneficial, vital relationship with the natural systems that provide its resources (Naboni, et al., 2019). A regenerative city fosters a symbiotic, mutually beneficial connection with its hinterland by reducing its environmental impact and actively strengthening and regenerating the productive capacity of the ecosystems on which it depends (Foggin, Brombal, & Razmkhah, 2021; Craft, Ding, & Prasad, 2021).

This chapter tackles the case study of Heliopolis as a place designed to be a regenerative place and how it turned from being human-friendly to a vehicle-friendly city. This transformation could go against regenerative cities to be closer to degenerative. In 1905, Heliopolis was planned based on the European movement of garden cities, seeking a utopia where people could live together with nature in harmony (Dobrowolska & Dobrowolsky, 2007). The idea was to create a self-contained living place surrounded by mixed uses, greenbelts and wetlands while linked to other Cairo districts through railway networks. At the end of construction, this city can provide a person's daily needs, better opportunities, and quality of life (Abusaada & Elshater, 2021a). Heliopolis had achieved the concept of garden cities as convenient, liveable, and regenerative places, but not for a long time. Through time, Heliopolis has faced waves of dramatic changes in its social context, urban nature, and economic environment (Elshater, Abusaada, & Afifi, 2019).

Today, Heliopolis faces its latest wave of change to its urban nature by implementing a rapid urban infrastructure development and upgrade plan. Admittedly, some significant issues are remarkable, such as the loss of green areas and public spaces, spatial segregation, vehicle dependency, and human safety factors. However, the absence of proactive social, urban, environmental, and economic studies, the lack of transparency, and the fast-track implementation of this plan have accentuated the incomplete planning approach and the contradiction between leadership's vision and actions on the ground.

This chapter highlights the effect of the significant loss of green areas and increasing vehicle dependence in the Heliopolis context while introducing its predicted consequences and contributions to raising the heat island effect and climate change in Cairo in the long term. In addition, to understand some spatial changes and represent the missing placemaking qualities of some public places in Heliopolis due to unilateral decisions and social exclusion. Therefore, this research selected a site within Heliopolis as a sample of other areas that have been affected and reached by the recent urban development actions under the strategic development plan of Egypt's Vision 2030.

To overcome the research challenges and interpret the problem from different dimensions, the present research integrated data collection methods as a significant advantage in filling the gaps and finding rational answers. This chapter followed four-step procedures. The first used analysed and simulated micro-climate urban heat island effects via the ENVI-Met lite software version. Second, the researchers of the present chapter used tracing satellite images from Google Earth to compare the spatial changes of some sites in Heliopolis throughout different time intervals. Third, the researchers of the current work conducted online questionnaires to determine the level of social satisfaction, concerns, and attachment after completing the plan. Fourth, direct field observation revealed the changes in users' attitudes during a real-time spatial experience.

The present work contributes to our understanding of degenerative cities through incomplete urban development. Besides, this study lays the groundwork for future research into the decision-making and governance framework in the Egyptian context towards a more sustainable approach, theoretically and practically, and applicable on the ground.

## BACKGROUND

The meaning of the term "regenerative" refers to an act of improving a place or system by making it more active or successful (Cambridge, 2021). This term goes beyond sustainability in urban disciplines since it is not limited to the efficient use of natural resources and low carbon-emitting but boosts ecosystem services through technical and management solutions (Herbert Girardet/World Future Council, 2010). In terms of practicality, the term has started to be used in urban and environmental fields between 2008 and 2011 to describe a new strategy towards our present crucial challenges such as climate change, ecosystem deterioration, and carbon dioxide emissions (Hand, Weber, & Bluestone, 2017). Eventually, the term was coined to refer to the process of imagining metropolitan communities that are capable of meeting today's crucial urban concerns. Additionally, the term "regenerative cities" dates to the World Future Council's Commission on Cities and Climate Change and HafenCity University in Hamburg. In-depth, the term refers to a complete approach that cities may use to mitigate the industrial revolution's damage to our ecosystem services and climate change (Fayed, Elshater, & Rashed, 2020).

The idea of "regenerative cities" aims to build an effective interaction between cities and their hinterlands and natural resource regions to restore ecosystem conditions while minimising carbon dioxide emissions (Plessis & Brandon, 2015). To be more precise, this therapeutic partnership should be given additional economical, technical, and political options. The literature demonstrates that regenerative cities require comprehensive political, financial, and technological strategies that promote an environmentally beneficial relationship between cities and the ecosystems from which they derive their sustenance (Herbert Girardet/World Future Council, 2010). Additionally, UN-Habitat defines "urban regeneration" as a process of reclaiming underutilised assets and redistributing possibilities, therefore enhancing urban prosperity and quality of life (UN-Habitat, n.d.). In brief, the core idea behind the concept of regenerative cities is to design a city that functionally meets the needs of all residents while considering all ecological and economic aspects. In other words, cities should be prepared to become a part of the implemented solutions that face environmental and climate problems (Herbert Girardet/World Future Council, 2010).

In 1905 and 1906, Heliopolis (Misr El Gidida in Arabic) was established by the Heliopolis oases company, which Belgian Baron Édouard Louis Joseph Empain founded. The district was designed to become a suburb outside Cairo and considered the new urban expansion toward the eastern desert. In Heliopolis, the city has been designed to be a regenerated and self-contained city based on the garden city planning concept, which prioritises the needs of residents, the quality of public life and healthy outdoor environments as the primary goals be achieved. The city of Paris inspired the planning of Heliopolis.

Heliopolis was built based on the "Garden City" concept in the desert to host around two hundred thousand inhabitants. Also, it consisted of two main separate oases with unique planning, urban fabric, and architectural characteristics to be linked together and with the centre of Cairo through fast electrical trams. These two different oases have been built where the nearest to Cairo is the luxuries, one for rich European people; the farthest one for lower-income Egyptian people and workers (Abdelaziz, 2021; Dobrowolska & Dobrowolsky, 2007; Ilbert, 1985; Tim, 1985).

From the beginning, It was such a lifetime dream for Empain himself to build the project with certain advantages; firstly, the project should be self-sufficient financially without any support from the government; secondly, the entire project should be made in the desert from scratch; lastly, the project considered as one of Empain's dreams and not to be as a result of the planning project. It was a generic success for Empain to provide urban expansion towards a pure place in the desert outside Cairo. The success was how affordable private-sector enterprises would provide cosy accommodations with the self-sufficient fund. Also, the project success in making a considerable revenue and its first building started to appear in 1908. During that time, Heliopolis became a lovely living place and faced a significant increase in its population rate. From that time, Heliopolis became considered as "one of the poles of development poles in Greater Cairo", and for instance; the population number started with almost 1000 inhabitants in 1909; 6210 inhabitants in 1915; 9200 in 1921 and reached 224,000 by 1924 (Ilbert, 1985).

Worth mentioning, the Garden City planning concept shows the circular-centric formation of the city plan in which the growth of the city from the centre where the public buildings and central park exist; to the circumference, which is ended by a circular railway line and attached with the mainline and surrounded by agriculture lands. In addition, six magnificent boulevards are connected through five main rings "avenues" that divide the city into six equal curricular parts. The roads through the boulevards lead to the centre; meanwhile, the trees alien the avenues are fronting, looking to the Crystal Palace, and framing the city's visual image (Elshater, Abusaada, & Afifi, What makes livable cities of today alike? Revisiting the criterion of singularity through two case studies, 2019). The appearance of Heliopolis as a pleasant place of living remained for a long time until some urban development interventions have applied and tended to make changes in its urban fabric, social structure and green infrastructure. Moreover, Heliopolis is considered a perfect example of losing the city's identity under the claim of making "urban upgrade actions" through many political conflicts and social and economic pressure. The city faced the first changing move when it turned into an extroverted "colonial character" place for all people and not limited to a specific class. As it was known as an "indigenous city" and introverted only for Heliopolitans, some areas and activities had become accessible to other types. Accordingly, some urban spaces and land-uses had to be re-adapted where nobody has become excluded anymore (Abusaada & Elshater, 2021).

Nowadays, Heliopolis faces its second dramatic change when the decision-makers decided to reshape the streetscape and road networks, implement vehicles and pedestrian bridges as well as remove some transitional traffic areas. Indeed, these actions are supposed to impact environmental conditions directly, social place attachment, and people movement patterns (Hassan, Moustafa, & El-Fiki, 2019). The city can be viewed as a functional component of the nation's identity (Elshater, Abusaada, & Afifi, What makes livable cities of today alike? Revisiting the criterion of singularity through two case studies, 2019; Kirk, 2006; Selim, 2016). Admittedly, certain factors directly influence changing cities' identities worldwide. Especially the political powers and system of governance have a strong influence in forming the cities' morphology (Selim, 2016). What happened in Heliopolis was a direct political decision based on the leadership vision of Egypt 2030 to solve the massive traffic congestions, decrease the duration of the daily trip from east to west Cairo and enhance the connectivity between the New Capital City and old cities of Cairo.

*Figure 1. The case study location*

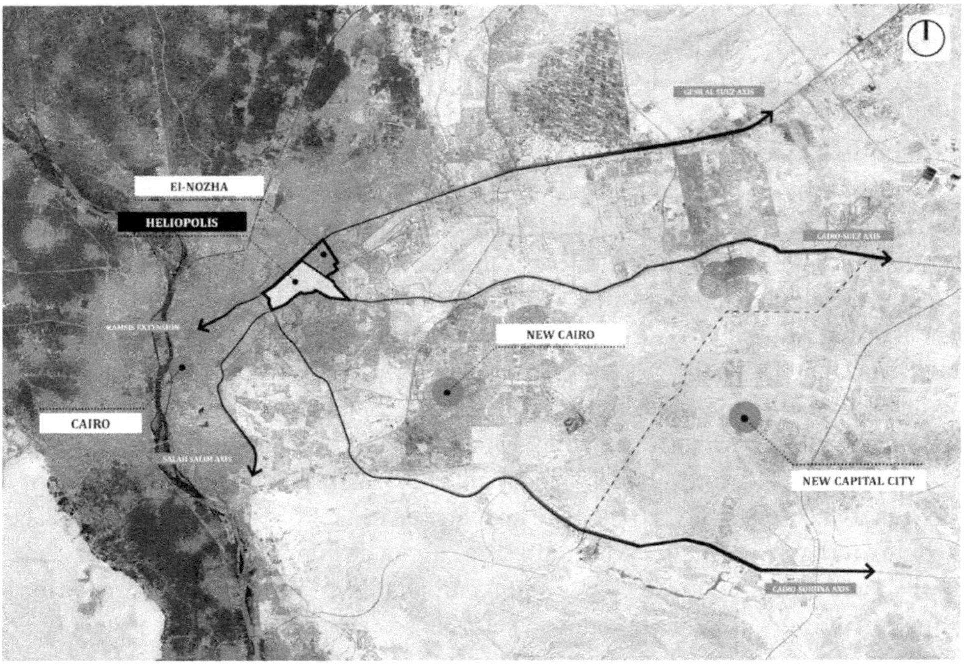


As shown in Figure 1, the target was to increase the capability by implementing new infrastructure decisions within the old cities in Cairo, such as Heliopolis, to accommodate the high traffic flow of vehicles from Cairo-Suez Cairo-Sokhna, Gesr Al Suze highways.

This integration will enrich the accessibility between the old settlements, new and futuristic projects and destinations towards the strategic development in the east direction. These objectives led to a significant increase in the usage of vehicles inside the city, and that what (Jacobs, 1961) pointed out in her book *"The Death and Life of Great American Cities"* how the dramatic increase in car traffic and the urban planning ideology of modernism that separates the uses of the city and emphasises individual free-standing buildings would put an end to urban space and city life and result in lifeless cities devoid of people. Also, Gehl and Rogers (2010) claimed in their book *"Cities for People"* that the modernist rejection of streets and the traditional city in the 1920s and 1930s and the introduction of functionalist ideals of hygienic, well-lit dwellings resulted in visions of the tall widespread town between freeways. Walking, cycling, and meeting others in shared urban spaces were not part of these visions, which profoundly impacted new urban development worldwide in subsequent decades. However, the Global Designing Cities Initiative

(2016) refers that people are shaped by their streets, and in our case, Heliopolis may undergo long-term changes, particularly on the social front.

## METHODS AND MATERIALS

Four main research methods were used to collect qualitative and quantitative data. The first trace and analyse satellite images of different sites in Heliopolis and surrounding areas to record differences before and after the implementation plan. Second, the field observation focuses on the implications for quality of life, place quality, and social aspects following place-making standards. Thirdly, after completing the development process, an online questionnaire was administered during two different time durations to reflect the social place attachment level. Finally, a pilot research project has been launched to simulate a selected site's microclimate and urban heat island based on specific parameters.

### Tracing Satellite Images of Heliopolis

Heliopolis is about 9.17 km$^{2,}$ with an estimated total population of about 139,856 inhabitants. According to the most recent official statistics on the government website (chart.cairo.gov.eg), the population consists of 70938 males and 68918 females. Heliopolis is divided into four central districts: Almaza, Mansheyet El-Bakry, El-Montaza, and El-Bostan (gopp.gov.eg/eg-map). Also, more than four road axes pass through and surround Heliopolis. These roads connect the district with the other communities, reflecting the magnitude of the daily heavy traffic flow within the district context. Figure 2 shows the location and the types of the implemented interventions within the context of Heliopolis and El-Nozha Districts. This site of Heliopolis was selected for deeper investigation.

The scope of the study border includes a small part of the El-Nozha district, which is considered locally as the new extension of Heliopolis but in another administrative division. Indeed, some popular spaces, such as squares, roundabouts, and street parks within this small part, have been affected and lost large green areas. Also, the choice of "Abo Bakr El Sedeeq Street" as a collector street is one of the most famous streets in eastern Cairo that has witnessed many radical changes. Despite the importance of this street in the vehicle infrastructure and mobility network, it has become a segregator and an edge between the two residential districts.

Previously, Abo Bakr El Sedeeq Street was considered one of the most beautiful street avenues in Heliopolis, which has lots of trees and palms and a large grass area. Also, the street was an excellent example of the streetscape standards in having a perfect distribution of the movement pattern and space management. The road

*Figure 2. The locations of the implemented interventions and selected site for investigation*

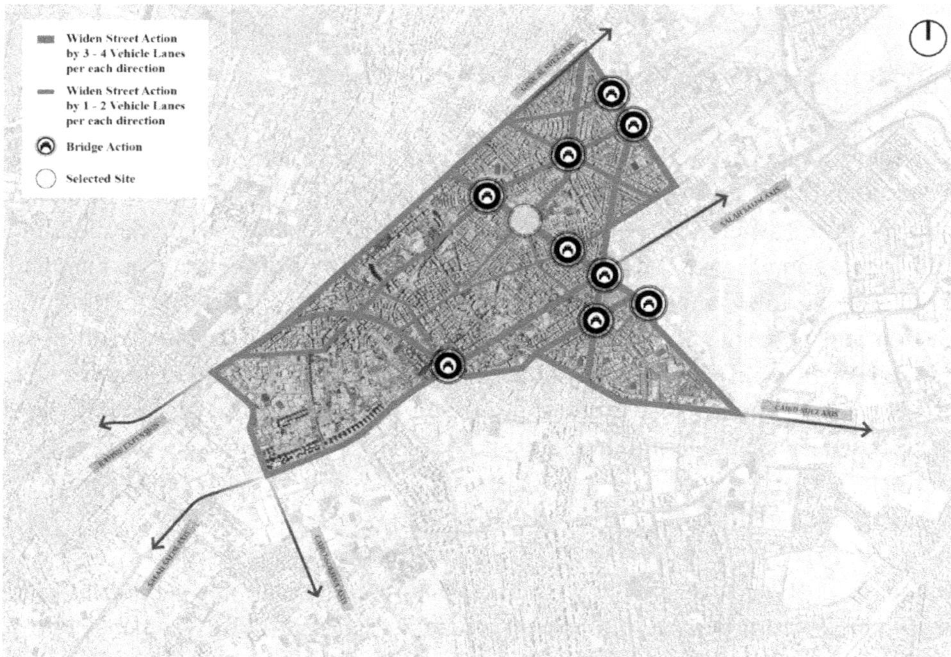


served commercial, educational, social, recreational, and mobility services on both sides, with precise control of the sidewalk, free zone, and traffic zone. Besides, the existence of the middle refugee island in the middle of the street provided the tramline pathway and its platform, which combined with the natural elements of the district. On the other hand, the more daily activities have increased, the more traffic problems have appeared and become complicated over time.

## Field Observation

The field observations aimed to discover the social side, detect different spatial changes on the ground, and highlight their consequences on Heliopolis's urban nature and the behaviour of other user groups. Some unfamiliar evidence of this area has popped up after completing the latest urban development plan. However, the observation was based on the diversity of the daily personal experiences in the Heliopolis context. It was essential to understand more about the intersected movement patterns for different mobility and walkability layers, the lost space quality, and unwelcome social issues. Hence, a pilot project has been created to

*Figure 3. The zone of the field observation*

be a sample of some other places in Heliopolis and El-Nozha districts that have been changed after the development's completion. Additionally, the methodology of the field observation was to record the change in people's attitudes and spatial experiences within the selected site after the completion of the development while comparing the findings to the previous experience of the same place before the start of the development producers.

After completing the development works, the first field observation took place on July 15, 2020. The field scope was within a circumference radius of 100 meters. It was in a standing alone position for 40 minutes in the afternoon at different locations within the selected site, as shown in Figure 3. The aim was to highlight the spatial changes, unfamiliar social evidence, and unwelcoming consequences of the place. Thus, some notes have been taken to record the users' experiences, the noticeable loss of place qualities, human behaviour, and the ongoing problem, followed by the predicted consequences of the urban and environmental perceptions.

The second field observation has been conducted between September 10 and December 29, 2020, for more than three months in the form of regular site visits with a minimum of three per week, or approx. Besides, 50–70 trips on different days, occasions, and statuses. This field observation has been made in two other states: driving a car and standing alone for 15–25 minutes. The aim was to observe the continued problems from the first observation and assess the responses and solutions made by the decision-makers regarding the street design failure compared to the first design of the street and referencing the standards of the Global Street Design Guide (GSDG) (2016).

The simulation of this method has been made through a digital built-up model for the selected site in Heliopolis that has specific social, environmental, and spatial parameters to simulate the radical spatial changes and connect their reflections on the human behaviour pattern and ecological dimension. Some digital software tools have been used, such as Cadmapper, AutoCAD, and Autodesk 3D Studio Max, to build the site context monochrome model while using other software such as Adobe InDesign and Google Earth Pro to simulate and create some illustrations, as well as highlight the detected findings of the field observation methods.

## Online Delphi Survey

This method is beneficial in studying the consequences of incomplete urban development and its impact on the social dimension in the Heliopolis area. Therefore, some collected answers to specific related questions have been discussed to reach a deep understanding of the social impact. These questionnaires have been selected from two online questionnaires made and submitted on two different occasions. The first one was launched before and during the executive development works, while the second one was after the completion and operation of the development plan for Heliopolis and El-Nozha districts.

As it was essential to know how the executed urban development plan follows the Egypt 2030 vision of sustainable urban development, an assessment of social satisfaction, community engagement and inclusion, and public participation was needed to evaluate the general approach of the development process and its consequences on people's lives. The benefit of this approach is that it enables us to investigate the anticipated and unpredicted effects of place attachment level and social belonging. Also, qualitative, and quantitative data offer a deep exploration of the influence of the implemented spatial changes generally and in some places in both districts, highlighting the similarities and differences in human place perception between before and after status.

Both two rounds of questionnaires were created on Google Forms, shared on Facebook, and sent by email to a total of 205 participants. The participants had submitted their answers for two rounds of surveys. The first survey had 17 primary questions and was conducted in both Arabic and English in November 2019 and reached 60 people. The second round of surveys had 16 questions and was conducted in the Arabic language and translated by the authors of the current work in February

2021. The total response from the two rounds reached 145 people. The time difference was essential to understand how users identified the problem and anticipated the consequences from the beginning. Also, the investigation of the social interaction and public response toward the changes on two different occasions helped to find the failure of the implemented development plan.

The first round of questions included:

1. Are you for or against making some developments in the Heliopolis area?
2. Do you expect more improvements from the action that is happening now?
3. Are you attached to the areas surrounding your home within the Heliopolis area?
4. Did anybody ask for your opinions or suggestions regarding this development as a resident in this area before the work started?

The second round of questions that were addressed to the residents after the development plans include:

1. Are you happy with the results of the development process in the region in general?
2. How would you describe the level of change in the region after the development process?
3. Do you feel safe for your family members or neighbours if they are on the street during daily hours?
4. From your point of view, has the Heliopolis area become more suitable for living after the development process?

## Micro-Climate Simulation and Envi-Met

In conducting the effect of urban heat islands based on the transformation in the case study, micro-climate simulation, and ENVI-Met were used. ENVI-Met is the state-of-the-art high-resolution microclimate modelling system. It is also a three-dimensional microclimate model designed to simulate the surface-plant-air interactions in an urban environment with a typical resolution of down to 0.5 m in space and 1–5 seconds in time. A significant advantage of using ENVI-met® software is that it enables users to make precise calculations and simulate and model a specific urban site (Elbardisy, Salheen, & Fahmy, 2021; Yang, Hu, & Marvin, 2021).

ENVI-Met is a predictive model based on the fundamental laws of fluid dynamics and thermodynamics. The model simulates flow around and between buildings, exchange processes at the ground surface and building walls, building physics, vegetation's impact on the local microclimate, bioclimatology, and pollutant dispersion. With ENVI-Met, we simulated the urban microclimate is an interactive system consisting of dozens of dynamic subsystems ranging from atmospheric dynamics, over soil physics, vegetation response, and down to building indoor climate. Typical areas of application are architecture, landscape architecture, building design, or environmental planning, just to name a few. From soil hydrology to building energy simulation, all systems are calculated in one big model, e.g., an urban quarter, allowing them to interact and adapt as an entire environmental system does. The ENVI-Met system can provide high-resolution data for any of these components, be it one building out of 500 or one trees within 1.500 trees (Ambrosini, Galli, Mancini, Nardi, & Sfarra, 2014; ENVI-Met, 2021).

This study adapted the limited lite-version of ENVI-Met software used to assess and evaluate only the impact of street vegetation on the local microclimate of our selected site in Heliopolis using the correct site coordination (30° 6.27 "N, 31° 19.58 "E). This version offers a limited model size of 50*50*40 (x-y-z), with a maximum length of grid cell of 2 m (dx-dy-dz) and limited simulation options. The selected measures of the simulation are to meet the exact summer dates of June 30, 2020, and June 30, 2021 with the same parameters of midday conditions at 12:00:00 pm, which meet the peak of sun exposure.

# RESULTS

## Satellite Images Investigation

Figure 4 depicts an example of a radical transformation that occurred in one of the most famous streets in Heliopolis and El-Nozha Districts, Abo Bakr El Sedeeq Street, between 2010 and 2021. This street is about 2.10 km long with a wide green refugee island in the middle, which creates a green avenue and contains the tram line and its transit stations. This street is located as a collector on Heliopolis and El-Nozha Districts' border. Previously, the street was well designed and managed to suit the existing highly dense movement patterns such as pedestrians, cyclists, tram lines, and vehicles. After the development completion, the street becomes a wide grey slot and is considered an urban spatial edge. The street is designed for the vehicle's movements only with the discarding of human and urban aspects.

## Field Observation Results

Undoubtedly, these spatial changes have affected the movement pattern, changing the shape and adding new, unfamiliar movements from both sides of the residential area. Figure 5 shows that changes happened at these two sites during 2010–2021.

*Figure 4. The transformation in the selected site from 2010 to 2021*

Many pedestrians have only one access point rather than two in the past. This access was not reachable for all users. It becomes clear that one entry cannot compromise the high pedestrian flow from both sides at all times of the day. Both genders face difficulties crossing the street, and especially the elderly, over 50, have to wait 10 minutes on average to cross the street on foot, compared to 5–10 minutes for young people or family groups, and 2–5 minutes on average for young people in a hurry to move or run. Due to the failure of the applied traffic system, there is a significant risk while crossing the street, in which they have recorded many pedestrian accidents and car crashes due to certain factors.

*Figure 5. The change in movement patterns between 2010–2021*

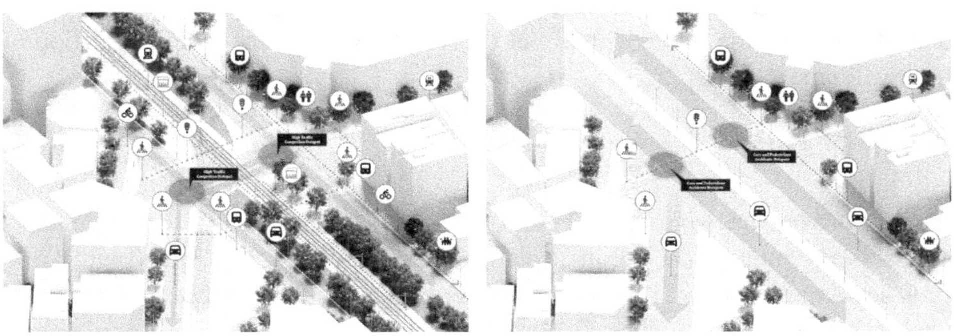

The appearance of this place has changed, with a substantial loss in its spatial values and quality. Some social and environmental activities have been observed over the past years before the latest urban development in this place. As illustrated in Figure 6, around 16 different activities were sampled as samples of some of the other activities usually announced daily in this area. The activities at that place have

*Figure 6. The change in place qualities and values between 2010–2021*

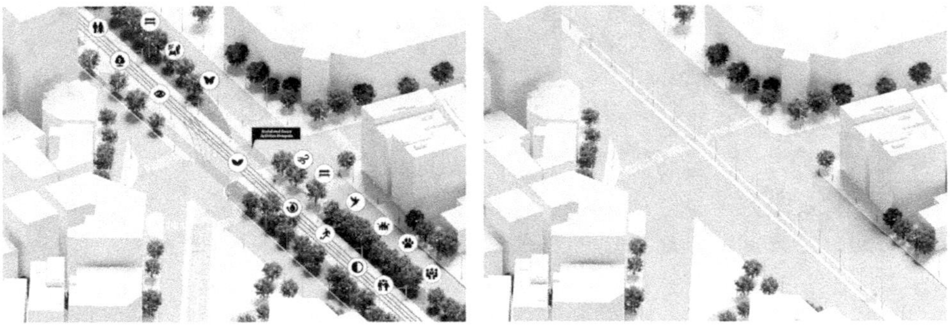

been shaped based on the user's interests and the opportunities provided by the original design of the street. The green part of this transit area had a significant role regardless of its simplicity. Also, the natural elements disappear, the tramline and stations are removed, and the refugee "transit area" becomes very narrow. It loses its quality, and the street becomes wider than before. The decrease in safe crossing lines and the absence of a fixed traffic system for pedestrians increase the risk of traffic fatalities and pedestrian accidents. Only one pedestrian crossway exists, which is not enough to cross the street legally with 50 metres of width through an unstable traffic system. Unfortunately, the system sometimes does not work and increases the risk of crossing. Due to the poor safety factor and the risk of accidents, some residents and visitors have begun to rely on vehicles rather than walking or cycling to avoid crossing the street. The use of cars has increased since the completion of the development.

## Delphi Results

The results from the two rounds of the survey showed that 63.1% of the participants are against the intended development plan in Heliopolis, while 21% agreed with doing some development work. 15.7% were undecided but partially agreed with the other two options in the same context. In the second question, this study asked participants if they expect more improvements from the development that is happening now. The answer indicated that most participants are pessimistic and do not anticipate more improvement in their living conditions. Around 32% expect nothing to be improved, and 39% see the possibility of more enhancement within their living place.

The survey participants' third question about their relation to Heliopolis indicated that approximately 98.24% of the participants were away from official public participation calls. In comparison, only 1.78% got a direct ask regarding the upcoming development action. However, around 29% are optimising to gain a better situation after completing the development actions. The results of the fourth question showed that 83% of the participants are attached to their living place in Heliopolis while 10% might be naturally connected. Besides, 7% are not linked to their surroundings within their living area.

The following are results yielded from the second group of questions in the second phase of the Delphi technique. The first question in this group indicated that 71% of the participants were unhappy with the new changes, while 29% were happy with the recent upgrade of the area. Indeed, this result proves that most residents, visitors, and place lovers understand that the implemented upgrading actions come before their social and urban needs. In the same line of receiving a high percentage, the second question figured out that 52% of the participants describe the level of these changes as radical changes, 46% tell them as considerable, 4.9% for moderate, and 0.7% of the participants consider it a superficial level of evolution.

When asking participants about their safe feelings, 49.7% of the participants are still attached to and love their living place in Heliopolis regardless of the unwelcoming urban changes. On the other hand, 45.4% of the participants started to lose their feelings towards their places, while 4.9% of them became careless about the changes in their places. The last question in this group of respondents mentioned that local streets within the Heliopolis area have become unsafe for pedestrians, where 76.8% of the participants do not feel safe for themselves, family members, or even neighbours during their daily trips due to the increasing rate of crossing accidents and vehicle crashes as well as the unstable traffic system. While 23.2% of them can feel safe and enjoy their outdoor journeys at any time of day.

## Results of Environmental Mongering

In the present case studies, the use of ENVI-Met software was specified and limited to assessing and evaluating only the impact of street vegetation on the local microclimate of our selected site in Heliopolis in two different situations. Also, this method completes and integrates with the previous techniques to fulfil the explanation regarding the anticipated environmental consequences, especially the urban heat islands (UHI), due to the massive divestiture of the green area and natural elements within the Heliopolis context. Additionally, this method is followed by a comparative analysis of the situation. Figure 9 shows cases before [A] and after [B] constructing bridges. The results confirmed that the microclimate conditions changed after implementing urban development. Surface temperature, $CO_2$ emissions, air quality, and thermal comfort have all been simulated to identify the significant environmental concerns and impacts on the place's residents and users.

*Figure 7. The survey results of the first round*

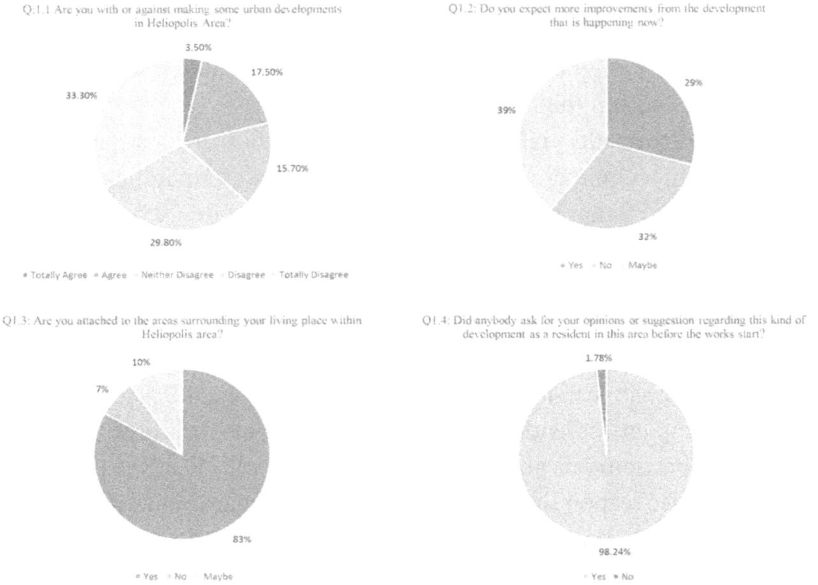


*Figure 8. The survey results of the second round*

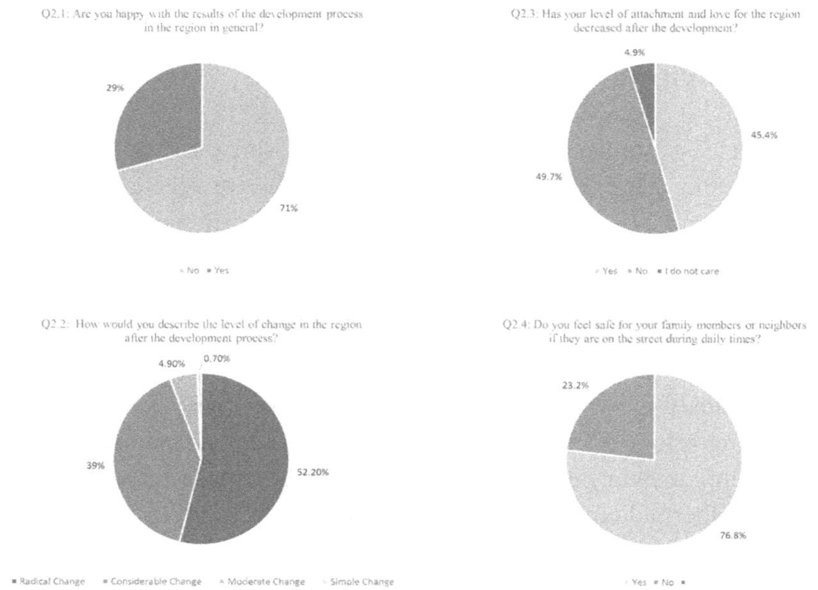

*Figure 9. The changes in climate condition after widening streets and constructing bridges*

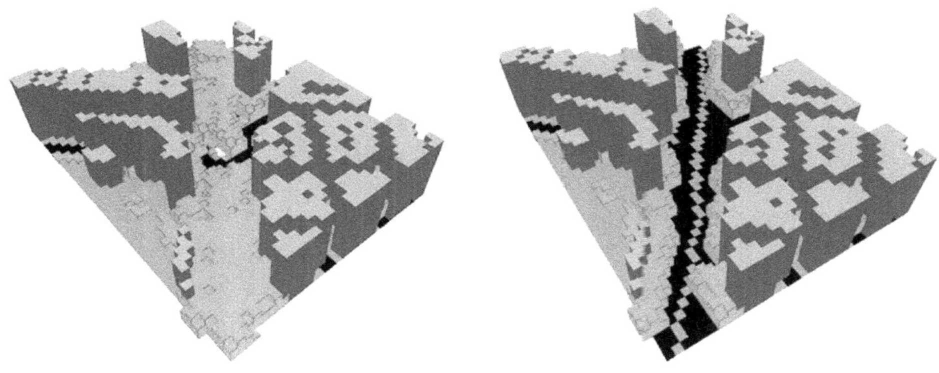

(A) (B)

As shown in Figure 10, the touchable impact of losing street vegetation and the natural element of this place tends to increase the temperature by 1 ° to 2 °C. The heated maps show the significant change in street temperature by adding more asphalt and removing ground covers and evergreen trees. The place has lost large canopies of trees that help cool down the air, enrich the comfort conditions, and purify the area's micro-climate. The deep red colour refers to 36–38 °C in some places, while the violet colour refers to a temperature of 38–40 °C or above that

*Figure 10. The impact of removing the street vegetation on air temperature*

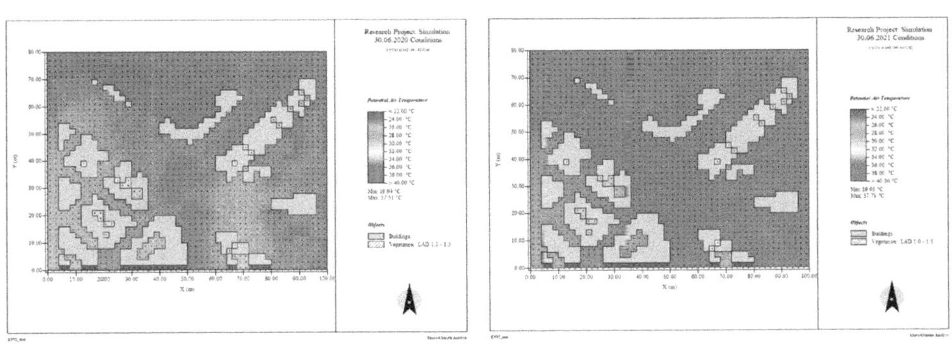

(A) (B)

*Figure 11. The impact of removing the street vegetation on air quality*

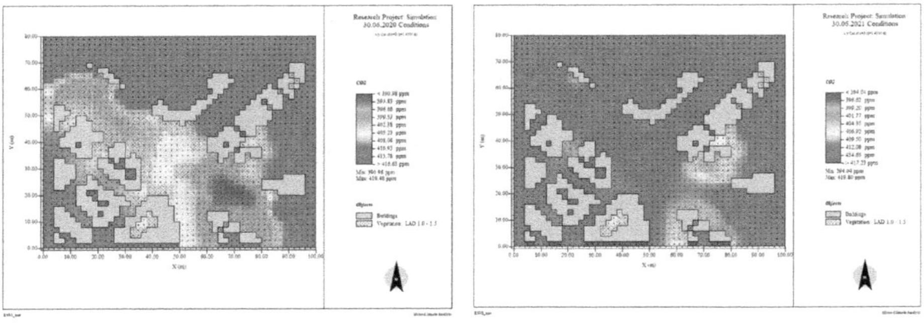

(A) (B)

appears in other areas. This significant loss of street vegetation directly influences the air quality of the place.

The presence of trees and green covers reduces the carbon dioxide amount in the air. It increases the oxygen, which directly improves the quality of urban life and living conditions in the area. The results represented the remarkable differences in the $CO_2$ amount after losing the natural elements of the street. Map (A) in Figure 11 shows that the $CO_2$ amount was lower than 390 and up to 408 ppm before the loss of the natural area, while map (B) shows significant changes in the $CO_2$ amount to reach 412 ppm or more at the same place.

Figure 12 shows the change in the wind speed in this area, which directly affects the human thermal comfort of the place. The street has become exposed to higher wind speed than before due to the loss of the vertical wind brake elements such as trees and palms. In its previous status, the wind speed has reached between 0.35 to 0.53 m/s rather than 0.18 m/s.

## DISCUSSION

A significant advantage of integrating four different methods is that it bridges the gaps and limitations of each one to achieve a comprehensive exploration of the core problem. Tracing satellite images and field observations came up with clear findings that show the dramatic spatial changes that have critical effects on the social aspects, which are reflected in the online questionnaire results. Meanwhile, simulating the environmental conditions shows the predicted consequences. Indeed, the site location of the case study is characterised by the diversity of residential,

*Figure 12. The effect of removing the street wind speed*

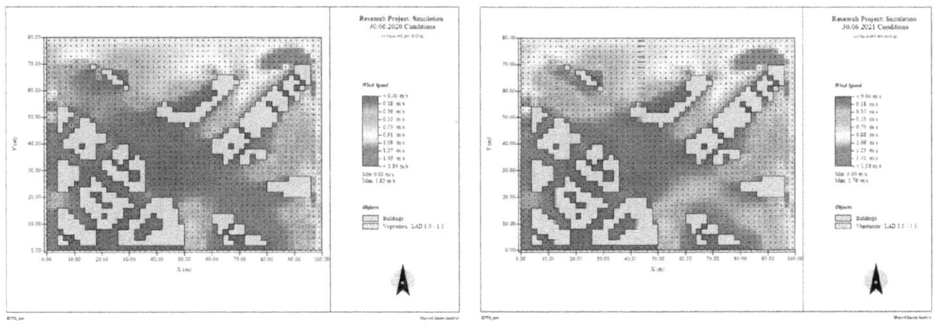

(A) (B)

commercial, transportation, educational, environmental, and recreational facilities and services with a very high volume of pedestrians and crossing street pedestrians. Also, it was interesting to explore the overlap between different movement patterns, such as tramline movement, vehicles, and pedestrians.

The main finding concludes that after completing development actions in different sites within Heliopolis and El-Nozha districts, admittedly, some sites have become scorching spots for vehicle and car accidents. Also, the site has become known as an unsafe and critical crossing area because of certain conditions. However, some solutions have been applied to limit and curb the accident rate, such as fixed traffic light systems, crossing zebra lines, vehicle radar, and pedestrian bridges.

The transformation of the urban pattern of the district creates notable spatial segregation because of widening the central and local streets, losing some transit islands, and implementing around ten massive concrete vehicle bridges. Some streets segregate the urban context and drive down the walkability and accessibility rates. Also, some places have lost their landscape and streetscape appearance and identity, which negatively impacts the urban heritage perception of the district and the community and environmental aspects. The result is consistent with the findings of the other methods. For instance, most of the questionnaire participants were against any changes to the development actions because of their predicted consequences on crossing the street, losing some natural elements and the urban spirit of their living place. It was not surprising that most of the participants described the changes as radical after losing great urban, social, and environmental values.

One anticipated and common finding is that the traffic flow becomes smoother upon entering or exiting the district, which helps shorten the duration of the daily trips between east and west Cairo and solve the massive traffic congestion within

Heliopolis some other places. Additionally, upgrading the infrastructure plan has noticeably succeeded in increasing connectivity between old districts in Cairo and the new administrative capital city. Besides, most people have started to face new difficulties during their daily outdoor trips, especially when crossing the streets in a scary situation. Also, it is interesting to note that the previous results match the observed findings that some places have lost their daily social activities and functions.

The finding also confirmed that some public places provided essential environmental and social values and qualities for all sites' users. The landscape part of the transit refugee island, for example, was beneficial for low-income people who had direct and free access to it; some social activities, such as gathering, resting, chilling, picnicking, playing, sleeping, grouping, and so on, were provided under a large canopy of trees, including an avenue of low wild grass and plants.

Moreover, natural elements, such as trees, shrubs, and grass, were beneficial ecologically for absorbing carbon dioxide and purifying the air by producing oxygen and enhancing the micro-ecosystem of a place within a highly dense urban context. Besides, this green spot has been used as the home of some native species, local birds, butterflies, and street pets such as dogs and cats. Meanwhile, it has contributed to managing the stormwater during the rainy seasons. For a long time, these added values and the sustainable contribution helped in maintaining the balance between the built environment and the ecological features for a long time.

The results from using ENVI-Met show that it is considered a significant loss of these previous values and ecological contributions after completing the urban development plan. It is anticipated that this considerable loss could influence the rise in surface and mean temperatures, carbon dioxide, and outdoor thermal comfort. This finding has been proved by simulating the micro-climate of the selected site, which shows the impact of removing street vegetation, which would strongly affect the quality of the outdoor spaces and environmental conditions.

The heat exposure analysis results are like the results from previous studies showing heat maps and thermal pattern associated with the UHI phenomenon (Elmarakby, Khalifa, Elshater, & Afifi, 2021; Elshater, et al., 2022; Yang, Hu, & Marvin, 2021). Our results from the Heliopolis case also go in line with the report by UN-Habit (2014). These methods result in the development process being shaped by its strategic vision, putting people at its centre. Before making spatial choices, many of the city's problems may be traced to a lack of thorough strategic planning (Elshater, Abusaada, & Afifi, What makes livable cities of today alike? Revisiting the criterion of singularity through two case studies, 2019). In Heliopolis' case, our results are on the same ground: previous research has shown that spatial planning should be connected to a holistic vision. If the community shares this vision, it enhances the planning process. Dreams of success include spatial dimensions that represent the city's distinctive cultural and physical characteristics; they guide the actions of all stakeholders, inspire them to work together, and guarantee that everyone is working toward the same objective.

## CONCLUSION

The purpose of the current study was to determine the reasons behind the dramatic change in Heliopolis and how it turned into a degenerative living place. Also, to explain how incorrect and incomplete urban development decisions could lead to a significant deterioration of the urban environment and green infrastructure, which directly influences the Urban Heat Island and has direct demerits on the quality and social life of the place. The findings of this study indicate that incomplete procedures can potentially turn the intended benefits, goals, and advantages of urban development actions into major disadvantages and failures. The results of this research support the idea that the unilateral approach of any development or upgrading plan that is oriented only to a specific side could lead to a magnificent failure on the other side of the development. This study has raised important questions about the nature of the sustainable urban development plan through Egypt's Vision 2030 and the contradictions found in the implemented actions. This study also contributes to our understanding of initiating the UHI and the consequences of social exclusion due to the incomplete urban development plan.

ENVI-Met simulation may be unable to correct building heights, materials, and traffic flow rates within the selected site. Furthermore, the sample size of the field observation was not recorded correctly, especially the number of people crossing the street, which affected the accuracy of the estimated crossing time per user type. Despite those limitations, the study certainly adds to our understanding of the importance of social and environmental dimensions in sustainable urban development. Therefore, this would be a fruitful area for further work to explore the contradiction found between the intended goals of the development plan and what has been implemented. Also, future studies are required to establish the viability of the built-up simulation model by using the latest version and more advanced features of ENVI-Met, while increasing the sample size of people during the field observation, which should be a focus on how local streets have been turned into urban edges and segregation within a highly dense urban context.

## REFERENCES

Abdelaziz, L. (2021). Heliopolis city, a cultural political interpretation. *AGORA CAUMME 2020-Mediterranean: Between expansion and regeneration.*

Abusaada, H., & Elshater, A. (2021). Effect of people on placemaking and affective atmospheres in city streets. *Ain Shams Engineering Journal, 12*(3), 3389–3403. doi:10.1016/j.asej.2021.04.019

Abusaada, H., & Elshater, A. (2021a). Improving visitor satisfaction in Egypt's Heliopolis historical district. *Journal of Engineering and Applied Sciences (Asian Research Publishing Network), 68*(1), 1–22. doi:10.118644147-021-00022-y

Abusaada, H., & Elshater, A. (2021b). Revealing distinguishing factors between Space and Place in urban design literature. *Journal of Urban Design, 26*(3), 319–340. doi:10.1080/13574809.2020.1832887

Ambrosini, D., Galli, G., Mancini, B., Nardi, I., & Sfarra, S. (2014). Evaluating mitigation effects of urban heat islands in a historical small center with the ENVI-Met® climate model. *Sustainability, 6*(10), 7013–7029. doi:10.3390u6107013

Cambridge. (2021). Retrieved 2022, from Cambridge.com: https://dictionary.cambridge.org/dictionary/english/regenerative#:~:text=regenerative%20adjective%20(GROWING)&text=relating%20to%20something%20growing%20or,powers%20of%20fire%20on%20cropland

Craft, W., Ding, L., & Prasad, D. (2021). Developing a decision-making framework for regenerative precinct development. *Sustainability, 13*(22), 12604. doi:10.3390u132212604

Dobrowolska, A., & Dobrowolsky, J. (2007). *Heliopolis rebirth of the city of the sun*. American University in Cairo Press.

Dovey, K., & Pafka, E. (2017). What is functional mix? An assemblage approach. *Planning Theory & Practice, 18*(2), 24–9267. doi:10.1080/14649357.2017.1281996

Elbardisy, W. M., Salheen, M. A., & Fahmy, M. (2021). Solar irradiance reduction using optimized green infrastructure in arid hot regions: A case study in El-Nozha District, Cairo, Egypt. *Sustainability, 13*(14), 9617. Advance online publication. doi:10.3390u13179617

Elmarakby, E., Khalifa, M., Elshater, A., & Afifi, S. (2021). Tailored methods for mapping urban heat islands in Greater Cairo Region. *Ain Shams Engineering Journal, 13*(2), 1–12. doi:10.1016/j.asej.2021.06.030

Elshater, A., Abusaada, H., & Afifi, S. (2019). What makes livable cities of today alike? Revisiting the criterion of singularity through two case studies. *Cities (London, England), 92*, 273–291. doi:10.1016/j.cities.2019.04.008

Elshater, A., Abusaada, H., Alfiky, A., El-Bardisy, N., Elmarakby, E., & Grant, S. (2022). Workers' satisfaction vis-à-vis environmental and socio-morphological aspects for sustainability and decent work. *Sustainability*, *14*(3), 1699. Advance online publication. doi:10.3390u14031699

ENVI-Met. (2021, October 21). *ENVI-MET: A holistic microclimate model*. Retrieved December 25, 2021, from Envi-Met: https://envi-met.info/doku.php?id=root:start

Fayed, L., Elshater, A., & Rashed, R. (2020). Aspects of regenerative cities. In S. Kamel, H. Sabry, G. F. Hassan, M. R. Ahmed, S. A. Elrahman, D. K. Hassan, & R. Rashed (Eds.), *Architecture and Urbanism: A Smart Outlook* (pp. 303–319). Springer. doi:10.1007/978-3-030-52584-2_22

Foggin, J. M., Brombal, D., & Razmkhah, A. (2021). Thinking like a mountain: Exploring the potential of relational approaches for transformative nature conservation. *Sustainability*, *13*(22), 12884. doi:10.3390u132212884

Gehl, J., & Rogers, L. R. (2010). *Cities for People*. Island Press.

Global Designing Cities Initiative. (2016). *Global Street Design Guide*. Island Press.

Hand, G. H., Weber, R., & Bluestone, N. (2017). Regenerative cities: Moving beyond sustainability–a Los Angeles case study. *Journal of Urban Design and Mental Health*, *3*(4).

Hassan, D. M., Moustafa, Y. M., & El-Fiki, S. M. (2019). Ground-floor façade design and staying activity patterns on the sidewalk: A case study in the Korba area of Heliopolis, Cairo, Egypt. *Ain Shams Engineering Journal*, *10*(3), 453–461. doi:10.1016/j.asej.2018.12.006

Herbert Girardet/World Future Council. (2010). *Regenerative Cities*. Hafen University.

Ilbert, R. (1985). Heliopolis: Colonial enterprise and town planning success? In A. Evin (Ed.), *The Expanding Metropolis: Coping with the Urban Growth of Cairo* (pp. 11–15). Concept Media/Aga Khan Award for Architecture.

Jacobs, J. (1961). *The death and life of great American cities*. Vintage Book.

Kirk, T. (2006). Framing St. Peter's: Urban planning in Fascist Rome. *The Art Bulletin*, *88*(4), 756–776. doi:10.1080/00043079.2005.10786317

Moore, G. M. (2020). Connectivity of public open space: Its meaning for different functions. *Journal of Urban Design*, *26*(3), 279–295. doi:10.1080/13574809.2020.1801340

Naboni, E., Natanian, J., Brizzi, G., Florio, P., Chokhachian, A., Galanos, T., & Rastogi, P. (2019). A digital workflow to quantify regenerative urban design in the context of a changing climate. *Renewable and Sustainable Energy*, *113*, 1–15. doi:10.1016/j.rser.2019.109255

Plessis, C. D., & Brandon, P. (2015). An ecological worldview as basis for a regenerative sustainability paradigm for the built environment. *Journal of Cleaner Production*, *109*, 53–61. doi:10.1016/j.jclepro.2014.09.098

Selim, G. (2016). Between Order and Modernity: Resurgence Planning in Revolutionary Egypt. *Journal of Urban History*, *42*(1), 180–200. doi:10.1177/0096144214566980

Tim, W. (1985). Heliopolis: a colonial enterprise that became an urban success. In A. Evin (Ed.), *The expanding metropolis: Coping with the urban growth of Cairo* (pp. 11–15). Concept Media/Aga Khan Award for Architecture.

UN-Habitat. (2014). *Urban planning for city leaders*. UN-Habitat.

UN-Habitat. (n.d.). *Urban Regeneration*. Retrieved from UN-Habitat: https://unhabitat.org/topic/urban-regeneration

Vandenbussche, L. (2018). Mapping stakeholders' relating pathways in collaborative planning processes; A longitudinal case study of an urban regeneration partnership. *Planning Theory & Practice*, *19*(4), 534–557. doi:10.1080/14649357.2018.1508737

Yang, J., Hu, X., & Marvin, S. (2021). Verifying an ENVI-met simulation of the thermal environment of Yanzhong Square Park in Shanghai. *Urban Forestry & Urban Greening*, *66*, 1–12. doi:10.1016/j.ufug.2021.127384

# Chapter 8
# Ergonomic Design for Pedestrians' Safety in Regenerative Cities:
## Heliopolis, Cairo, Egypt

**Nardine El-Bardisy**
https://orcid.org/0000-0002-3339-7396
*Ain Shams University, Egypt*

**Alaa N. Rashed**
*Ain Shams Univeristy, Egypt*

**Eman Ahmed Elmeligy**
*Ain Shams University, Egypt*

**Hala Hussien Aboulnaga**
*Ain Shams Univeristy, Egypt*

## ABSTRACT

*Mobility and safety are the main characteristics of the city's urban strategies that regenerative cities adopt. This study is concerned with the mobility of pedestrians while crossing vehicle routes within urban developments. Several research groups recognise the negative-positive impact of ergonomic design in understanding human limitations and abilities; additionally, most researchers mentioned ergonomic design in the design of interior architecture elements, and a few have introduced it on an urban scale. So, research used ergonomic design to analyse pedestrian and vehicle movement. The research method adapted ergonomic design principles as a method to analyse human behaviour while crossing roads in an Egyptian context. After reviewing the relevant literature and Egyptian context, the authors discovered that studies show significant development occurred on Cairo's roads that threaten people's lives. The analysis tools used in research are questionnaires, surveys, and observations. The results concluded that both streets need more pedestrian-friendly designs and regulations.*

DOI: 10.4018/978-1-6684-2462-9.ch008

## INTRODUCTION

The process of creating public places, workplaces, goods, and systems in the city to fit people's wants and preferences is discussed in this chapter. We tackle the thoughts of ergonomics as a trend in city design based on people-oriented development (Occupational Safety and Health Administration, 1999; World Health Organization, 2021). In addition, seven characteristics of regenerative cities support the urban development method that attempts to form a beneficial relationship with the city ecosystem (World Health Organization, 2018). These characteristics are applied in urban strategies of cities, including mobility and accessibility, liveability, ecology, equity, Energy and Natural Resource Efficiency, resilience, and smartness (Hand, Weber, & Bluestone, 2017). Mobility and accessibility, cities increased their dependence on vehicles, which increased demand for fuel. Accordingly, the concept of regenerative cities focusses on shifting back to the movement of people to rebalance mobility options (Fayed, Elshater, & Rashed, 2020; Hassan, Rashed, & El Nagara, 2022). It is essential to design a suitable pedestrian network so focus on pedestrian movement while crossing roads through ergonomic design concept.

The ergonomic design of road carriageway and pedestrian cross paths became an essential factor in design because of drivers' and pedestrians' daily life experiences (Ahmed, et al., 2021; Anciaes & Jones, 2018; Chebanyuk, et al., 2020; Zafri, Rony, & Adri, 2019). The ergonomics field focuses on human abilities and limitations through deep investigation (Dohrmann Consulting, 2014; Eldar & Fisher-Gewirtzman, 2019). It is explained as the job to adapt workers' needs, not the inverse by the Occupational Safety and Health Administration (OSHA) (2022). Moreover, Tayyari and Smith (1997) consider it a branch of science that focuses on the human and work environment. While previous research defines it as an interrelation between human and systems, where the human is the focus, and the scope of ergonomics include cognitive, physical, and organisational aspects (Shoubi, Barough, & Rasoulijavaheri, 2013). This field seeks to improve people's interaction with the surrounding environment to reduce the risk of injury and harm (El Bardisy, Reuter, & Ismail, 2017; Elshater, et al., 2022; Shoubi, Barough, & Rasoulijavaheri, 2013). Research confirmed that road traffic injuries and death reduction are considered the main priorities in the design process due to the high death rate monitored globally (Bhalla & Gleason, 2020; Lin, et al., 2021).

Previous research addressed the challenge of creating ergonomic and convenient places for our daily lives (Ben-Bassat, et al., 2021; Ben-Bassat, Are ergonomically designed road signs more easily learned?, 2019), interior design furniture of home and workspace (Fusaro & Kang, 2021; Wang, Han, & Li, 2021; Saptaputra, Kurniawidjaja, Susilowati, & Pratomo, 2021). However, few studies discussed the feeling of human beings and their communication statically during road passing and safety, especially in the Egyptian context.

This research tackles the global problem related to road traffic which is a significant challenge. Reports document that 1.35 million people die yearly worldwide (World Health Organization, 2021). The global rate of accidents is high in African, American, and eastern Mediterranean cities (World Health Organization, 2022). When focusing on Egypt, the highest percentage of car death rate is between pedestrians and passengers, so they focus on the research (Naci, Chisholm, & Baker, 2009).

Shading lights on Egypt's traffic data monitoring show that the average death rate of accidents is around 10 per 1,000,000 population: the distribution of deaths throughout the years. As a result of drivers' and pedestrians' behaviour about the surrounding environment (Puvanachandra, et al., 2012). Indeed, it is essential to address the recent modification of the road network in Egypt; for instance, the road widening and new overpasses in Nasr City and Heliopolis in Cairo, the capital, and how it impacts the road accidents rate (ElMoghazi, 2019). While high-income cities are removing urban highways, they built only decades ago and replacing them with the new walking, cycling, and transit technologies.

This research tackles the problem of road accidents in some Egyptian sites because of the impact of road development on the behaviour of pedestrians while crossing roads and the response of drivers. This problem is reflected in the convenient atmosphere on highways and public places. In this context, this research focuses on these three aspects: road development, car accidents, driver, and pedestrian behaviour to deeply describe this correlation.

To clarify the research problems, this chapter analyses that change in correspondence to the behaviour of drivers and pedestrians and road accidents. In-depth, this chapter tackles the issue of how far the ergonomic design applies to a local case study in Egypt. Additionally, this research attempts to answer the following question: How does road development in Cairo impact pedestrian behaviour while crossing vehicular roads through ergonomic design methods?

The present chapter aims to present urban policy and design guidelines based on the findings yielded from the case studies. Moreover, the current work seeks to offer tools and methods to investigate the ergonomic design in roads and public places. The aim extends to evaluate two cases from Egypt, reaching lessons learned of future implications in cities of Global South. This article aims to revisit the design principles of this term and compare them with relevant terminology like universal design. The research objective is to eliminate/minimize car accidents through pedestrian behavioural analysis and understanding of ergonomics design. This investigation on the Egyptian case in Al-Hejaz, and the Gesr Suez Streets, can help urban developers and decision-makers in cities of the Global South to solve the research problems. In addressing the research aims, this research used qualitative study by surveying the proper methods to understand pedestrian behaviour while crossing the road, which is reflected in research as shown in Figure 1.

The motivation here came from the belief that people's lives are the most precious thing in the world. People are essential for the development of the economy and society. They make use of resources, and they, themselves, are also resources but with varying quality. Besides, the research output came from a deep motivation to achieve more safe, functional, and sustainable cities.

The contribution of the present study is in revisiting the categories, aspects and principles of ergonomic design that were first introduced in Western literature and needed to be adapted to the case in cities of the Global South. The novelty is in investigating the cases from Egypt that are rarely discussed in literature from that perspective. This motivation leads to a significant research contribution in the academic community. Thus, it positively impacts various scales as an integrated system.

*Figure 1. Research structure, source: the authors*

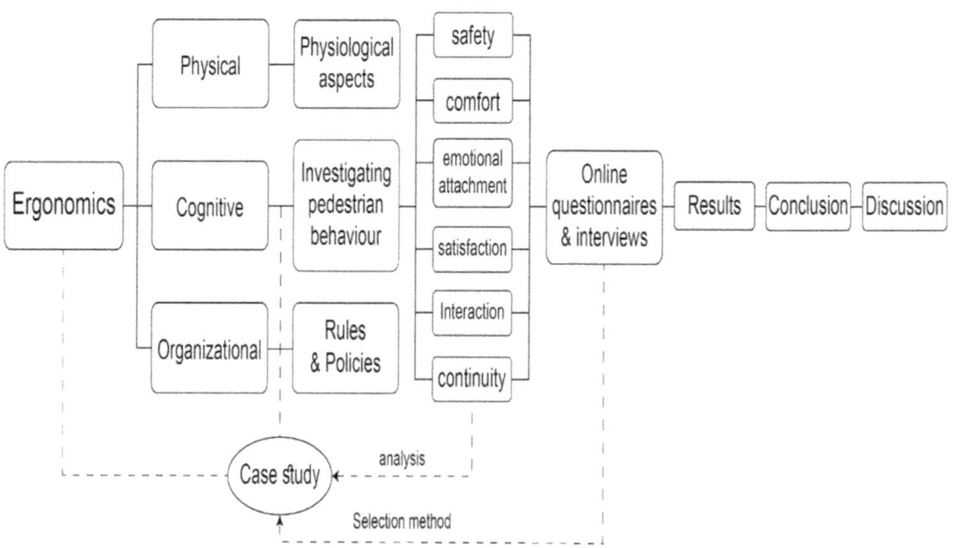


## LITERATURE REVIEW

This section gives an overview of the concept of ergonomic design and its principles and factors. The term ergonomics came from the Greek ergon word (work) and nomos word (law), which means detailed thought of laws inside work (Pao &

Kleiner, 2021). Polish scholar Ojciech Jastrzębowski first introduced the word in 1857 (Jastrzebowski, 1857). The term was known when he reprinted his book "an outline of ergonomics" in English, in 1997 (The International Ergonomics Association (IEA), 2022). Polish was focusing on the integration between work and health due to the political circumstances in japan at that time and the history of that relation in different ages (Occupational Safety and Health Administration, 2022).

The International Ergonomics Association (IEA) was founded based on various events; In 1949, the ergonomics research society (ERS) was established in England, which was the main reason for the IEA existence (Healthcare & Karwowski, 2006). The first meeting was held in 1961 in Stockholm. It explains the phases of the association and its activities. The IEA then expanded in 1967 to become the association of federated societies worldwide. Until 2011 when it became an international not-for-profit organisation in Zürich (Switzerland) under the official name as 'The International Ergonomics Association (IEA)' (2022).

Ergonomics is defined according to IEA as ergonomics (or human factors) as a scientific discipline concerned with the understanding of interactions among humans and other elements of a system, and the profession that applies theory, principles, data, and methods to design. This discipline aims to optimise human wellbeing and overall system performance. The concept 'human factors and ergonomics (HFE)' is used with the same weight or basic unit by IEA, as it is identified in the definition of ergonomics HFE is not a stand-alone (The International Ergonomics Association (IEA), 2022).

HFE is a user-centric integrating science. Its three main categories are physical, cognitive, and organizational, which various scholars use (Eilouti, 2021). These categories include perception, memory, reasoning, motor response, human-computer interaction, communication, participation, cooperation, environment, socio-techno systems, human anatomy, physiology, anthropometrics, and biomechanics (Consulting, 2014). HFE has three fundamental qualities; it uses a systems approach, is design-driven, and focuses on performance and well-being as its main goals (Consulting, 2014; Dul et al., 2012; Eilouti, 2021). Such a unique approach can address significant challenges regarding work, product, service systems, and technology adaptation (Dul, et al., 2012; Eilouti, 2009).

A holistic perspective is a primary consideration in ergonomics by using a systems approach. This systems approach considers the importance of a holistic view (International Labour Organization, 2021; Wilson, 2014). It studies the context and interactions among humans and their environment and understands the nature and complexity (International Labour Organization, 2021). Ergonomics applies theories, principles and knowledge from other related disciplines to the process of design and evaluation (Dohrmann Consulting, 2014).

Iterative processes such as the Plan, Do, Study, and Act (PDSA) cycle can be used in a systems approach and for continuous evaluation and refinement to ensure the validity and suitability of a system. It is essential that the design maintains a balance between human wellbeing and system performance (International Labour Organization, 2021).

HFE is concerned with the study and design of socio-technical systems (Stevens & Salmon, 2015). The research recommended that ergonomics is based on essential core values that are humans are the focus of the design, technology is a tool to serve humans, quality of life promotion, accepting individual differences, and stakeholders' responsibility (Read, Salmon, Goode, & Lenné, 2018). These values and principles should be involved in any ergonomic design (Eilouti, 2021).

Moreover, the science and practice of HFE are not domain-specific (Dohrmann Consulting, 2014). The field of 'urban design' has been explored through the lens of ergonomics in a few articles (Appolloni, Giretti, Corazza, & D'alessandro, 2020; Costa, Campos, & Villarouco, 2012; De Carvalho, 2012; Read, Salmon, Goode, & Lenné, 2018). An article issued in the *IEA 2012: 18th World Congress on Ergonomics* addressed the application of ergonomics on the city and how this links with sustainability and resilience, using the term 'Ergopolis' and defining it as *"an idea for thinking and transforming the city from the understanding of its complexity and the complex human activity in it."* (De Carvalho, 2012).

As presented in the previous literature review, many studies have investigated ergonomics on a narrow scale and the relationship between humans and their surrounding workplace environments, especially interior design standards and furniture design. However, few studies addressed the relationship between ergonomics and public spaces and how to integrate them with the pedestrian network, especially pedestrian crossovers in Egypt.

*Figure 2. Diagram showing how ergonomic design inspires the approach of this investigation*
*Source: the authors*

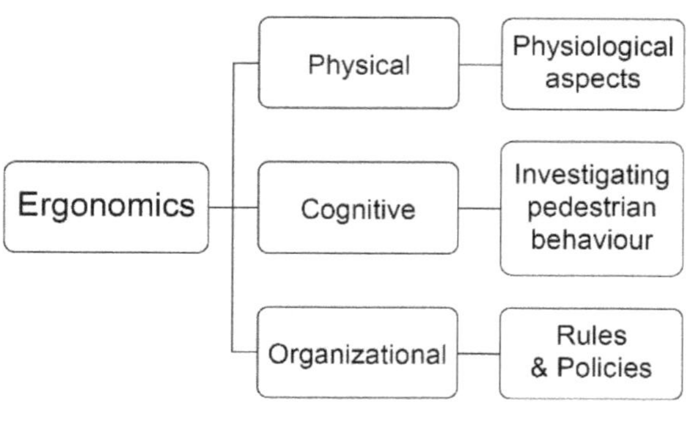

This chapter uses the science and practice of HFE to address the impact of the recent road network development on pedestrian mobility in the Egyptian context and achieve a suitable design that can minimize or eliminate road traffic injuries. Figure 2 illustrates a summary of ergonomic design, its three main categories, and how they are connected with the main focus. This figure looks into the six factors that literature and the nature of this case study say are essential: safety, comfort, emotional attachment, satisfaction, interaction, and continuity.

## MATERIALS AND METHODS

This section gives an overview of the case study selection and methodology of data collection in the case study of Heliopolis, Egypt.

### The Case Study

The eastern part of Cairo has received a large share of the development and expansion of roads. The upgrading and expanding of the roads increased the number of lanes from 3 to 4 in each direction, with a lack of adequate security and safety measures for pedestrian movement causing numerous accidents and deaths. They lack means of controlling the movement of vehicles and providing safety for pedestrians, e.g., industrial bumps to slow down traffic, traffic lights, bridges for pedestrians, and islands between the two directions of the road to provide a rest zone for pedestrians while street crossing. For instance, El Hejaz Street in Heliopolis, runs from Heliopolis Square to El Hejaz Square. Three lanes were added on each side after development. In addition, there are no pedestrian safety measures that were implemented after the change. Many other risky roads, such as Haroun El Rasid, El Mahkama, Farid Smeka, El Salam and Gesr El Suez streets.

This research carries out one case study, focusing on investigating cognitive as one of the ergonomic design categories and its principles. To select a specific case study, the Delphi method is used through a survey for users to choose clear rough two rounds. The first round of investigation was to select the two highest streets in Heliopolis to conduct the next investigation phase. The second phase included an extended focused survey exploring the challenges in creating Heliopolis' ergonomics.

### Data Collection

Using two rounds of Delphi methods as well as field observations, this study collected data qualitatively and quantitatively. Two rounds of Delphi methods were used to collect primary data through an online survey (Table 1). Specific roads in Heliopolis were the focus of the first round of the Delphi survey. The first round of investigation was to rate the threat level of those five roads: Haroun El Rashid, El Hejaz, Farid Semeika, El Salam and Gesr El Suez. The survey took place over seven days, and there were 43 respondents. It was conducted online. The survey results show that Gesr Suez and El Hejaz streets are the two most critical roads.

Following those results, a semi-structured online survey and observations were conducted as part of the second round of interviews. We deepened our relationship with users through this phase and studied their behaviour. In correlation with the literature review, these tools were applied to investigate pedestrian behaviour concerning the cognitive ergonomic category's principles (human). This second round of the online survey consisted of 13 questions about safety, comfort, emotional attachment and satisfaction, social interaction, and a general inquiry, as shown in Table 1. The authors collected various online, face-to-face, and phone data asking questions. The respondents' sample was a total of 11; seven responses were for El Hejaz Street and four of them for Gesr El Suez Street. Ten of the respondents' age group was 20–30, and a respondent was 30–50. Their genders were seven females and four males.

*Table 1. The survey questions*

| Delphi method | Items of investigation | Questions/ investigation |
|---|---|---|
| 1st round | Site selections for the second round | Rate the following streets based on the safety perspective from the lowest to the highest safe after the development changes. |
| 2nd round | Safety | Do you face any cross-border-related issues in connection with road safety? |
| | | Rate your feeling of safety level |
| | | In your opinion, what are the leading root causes of accidents during passing roads? |
| | Comfort | Do you feel comfortable while passing roads? |
| | | To what extent are you comfortable while crossing the road? |
| | | To what extent are you comfortable while crossing the road? |
| | Emotional attachment and satisfaction | Do you have any memories of this road? |
| | | To what extent are you satisfied with that road? |
| | Social interaction | How frequently do you attempt crossing this road? |
| | | What are the reasons that make you cross this road? |
| | | What are the activities that you would like to see in that street? |
| | General | Please state any recommendations for user experience improvement on that road. (Pedestrian crosswalks, pedestrian bridge, vegetation, … etc.) |

*Figure 3. Diagram of the research methods and tool, source: authors*

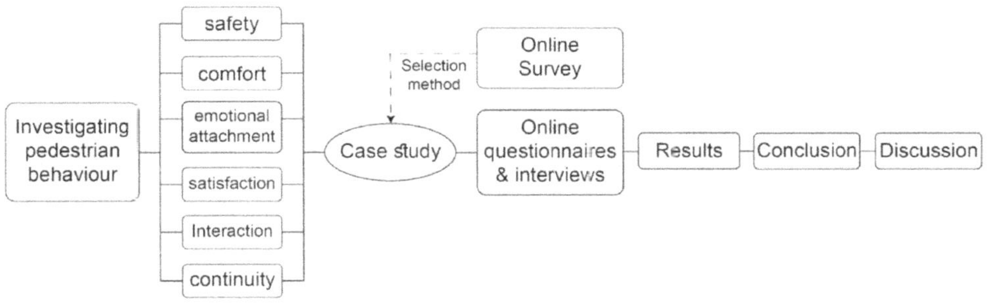


The observations took place late afternoon on Friday, December 17, 2021. The observations were made to investigate the structure of Gesr El Suez and el Hejaz Roads systems, such as a number of lanes, traffic signals, crosswalks, road signs and off crossing the street by pedestrians. Figure 3 shows a brief explanation of methods and tools.

## RESULTS AND DISCUSSION

This section provides an overview of the results of fieldwork, interlinked with a literature review. Starting with **El Hejaz Street**, the selection of the street included in the study area (from Heliopolis's metro stations to El Hejaz Bridge) was seen to have several six lanes on each side with a narrow crossing island, making it an extensive street for people to cross comfortably (see the following map of the study area). People were seen running and having trouble crossing to the other side. During the visit, a picture showed a group of kids running to reach the central island (see

*Figure 4. The map of El Hejaz study area (left) and a photo showing kids running to cross El Hejaz Street (right). Source: the authors and Google Earth, respectively*

Figure 4). To sum up, people think the street is dangerous and does not care about the safety or comfort of its pedestrians.

The results yielded from the case of **Gesr El Suez Street** show that the section of the street included in the study area (from Elshams Club to Huckstep Bridge) was seen to have different widths in different zones. These zones led to congestion in specific areas such as Alf Maskan, which lies in the zone of Elshams Club. The street has four lanes in each direction, then it becomes more expansive in the El Arab zone, having seven lanes in each direction with a large crossing island. Cars go very fast with no speed limit as there are neither radars nor cameras.

As shown in Figure 5 that was taken during the site visit, there are different types of activities, including a commercial zone on one side of the street with high traffic flow in one direction, and El Shams Club, on the other side, with low traffic flow leading to higher speeds in that direction—going to the El Arab zone with fewer activities and more residential zones.

*Figure 5. A map of Gesr El Suez study area and a picture of Gesr El Suez Street and its commercial activities, source: authors after Google Earth*

The survey results will provide more a precise and in-depth understanding of the different experiences and views of regular users of El Hejaz and Gesr El Suez Streets. This chapter conducted that there is a reflection on the widespread frustration and the need for change to a better design that can enhance the outdoor environment. People were eager to share information that might result in beneficial outcomes. The first part of the survey explored the significant concern of street safety. Participants were asked if they faced any issues while crossing the road about road safety. A majority of 85.7% and 100% answered with confirmation for El Hejaz and Gesr El Suez roads. When asking a street vendor in El Hejaz, the road conjunction helped people in several places across the street, especially elders who have trouble running. Interviewees were also asked to rate their feelings of safety from one to five (one being remarkably safe and five being extremely dangerous). For El Hejaz Street, five participants out of seven gave a rating of five, one showed a rating of four, and the last gave a rating of three.

For Gesr El Suez Street, four participants; rated their feeling of safety as five. The other gave a score of four. Another rated it as three, and the last one gave a rating of two. These first two questions show that widening these roads compromised the first safety attribute and that no measures were taken to mitigate this severe problem. When allowed to explain their answers, people revealed multiple issues and obstacles.

Respondents mentioned the roads being too broad with a large flow of speeding cars, making it extremely difficult for them to pass. Some complained that there are no dedicated crosswalks or bridges for pedestrians to use. Some added that there are no traffic lights, speed bumps or speed limiting measures to control the vehicles' speeds, giving a chance for people to cross. One explained having these safety issues by bringing up the critical problem of accidents. Others also confirmed this throughout the conversations by mentioning several accidents taking place. While interviewing the same vendor, he said there were road accidents on El Hejaz because of speeding cars, appley at night revealing another problem that concerns safety, ill street lighting.

A question explored the issue of accidents, asking participants from both streets to share their thoughts on the main reasons for this problem. Almost all the participants attributed this problem to the uncontrolled and un-provisioned high speed of vehicles. Some revisited the issues of street width, unavailability of dedicated pedestrian crosswalks and insufficient street lighting. A couple included the drivers' behaviour as being a cause; one talked about how some drivers quickly switch lanes and the other mentioned the drivers' control over their vehicles being a contributing factor. In summary, people's feelings of safety were hugely affected by the recent alterations and their consequences on the overall environment of both streets, which in turn affected comfort.

The second part of the survey discussed this issue. Participants started going more in-depth, explaining their feelings and unique experiences while interacting with the road and how it directly affects their comfort. It was clear from their replies to a question asking whether they feel comfortable crossing the street; for El Hejaz Street, about 85.7% confirmed their discomfort. The remaining 14.3% answered with 'maybe.' Furthermore, 100% of respondents were utterly uneasy on Gesr El Suez Street. When respondents were asked about the root cause of their discomfort, they mentioned feelings of scariness and fear, as car speed, lack of safe crossing methods contribute primarily to these feelings, which, in the end, is a reflection of safety. In addition, when they were asked to rate their comfort level from 1 to 5 (1 being extremely comfortable and 5 being extremely uncomfortable), 71.4% of the respondents rated 5, 14.3% rated 4, and the other 14.3% gave a rating of 3 for El Hejaz Street. For Gesr El Suez street, 50% of respondents rated it 5, 25% rated 4, and the remaining 25% gave it a 3. Lastly, when asked about their suggestions for solutions to the comfortability problem, their answers included pedestrian bridges, adequate lighting, speed bumps, traffic signals, which again reveal a direct reflection of their need for safety.

In contrast to the considerable safety and comfort problems, most respondents (85.7% and 50%) have memories of El Hejaz Street and Gesr El Suez, respectively, reflecting their emotional attachment for different reasons. Most of the memories reflected the happiness of moving between the commercial district and green spaces. Their schools sold products or walked with their dogs between trees with red flowers. The change that affected these memories causes colossal dissatisfaction.

On the one hand, when asked, 71.4% rated their satisfaction with El Hejaz Street atmospheres as 5, meaning completely dissatisfied, and the other 28.6% also shared similar feelings of discontent, placing it as 4. In addition, 50% were dissatisfied with Gesr El Suez Street, giving a rating of 5, and the other 50% were neutral and assigned a rating of 3. Respondents were asked about their frequency of crossing the road, and 42.9% pass El Hejaz Street daily, which is a usual hassle that causes problems for them. The other 57.1% pass El Hejaz Street a few times a week. On the other hand, 100% of Gesr El Suez respondents said they gave the street a few times a week. The reasons they mentioned for passing were parking, purchasing products, and using public transportation. Streets on both sides are very busy, but there aren't any safe places to cross them. This is because of the activities that happen on both sides.

## CONCLUSION

El Hejaz and Gesr El Suez streets are examples of dangerous roads in Cairo, Egypt. This research conducted a deep investigation of these roads focusing on the pedestrian crossing and exploring this activity through ergonomic design principles. From these analyses, a lack of safety is the leading cause for the problems in all the other factors (e.g. comfort, satisfaction, emotional attachment).

Out concluded remark confirmed the importance of safety to be tackled on different levels since it should be achieved on both governmental and individual levels. For the governmental level, regulations and policies should be settled to guarantee pedestrian safety and improve road design. For instance, road signs, speed bumps, pedestrian lanes, crosswalks, and pedestrian bridges are all crucial elements that should be used to improve the quality of the street for pedestrians and drivers equally. On an individual level, car drivers should follow regulations and have control over their cars. These are all simple and feasible solutions to achieve pedestrian safety and mitigate the effects that the recent changes have caused. Pedestrians are the most vulnerable to these changes, and no means have been incorporated in these two streets to ease the daily trips of their users.

The study has its limitations, especially in Egyptian case data collection. Although the authors of the present chapter chose only two case studies in Heliopolis due to time and data applicability, there are many updates in other regions of Egypt that can be analysed. The authors also faced a problem in encouraging people to respond to the survey. This study also faced the challenge of continuous updates to streets during data collection. Based on these limitations, this chapter suggests future research that can expand the data collection in other cases that witnesses changes in road structure that affect safety. Future research can also compare the present results with other similar topics.

## DATA AVAILABILITY

The data collection was designed to be collected in Arabic for better reachable information. The survey questions of the first round in the Delphi technique can be tracked using the link:

Questions link is: https://docs.google.com/forms/d/e/1FAIpQLSc6hRHBxpYbO54ORWBavoaSoOYbIS0fu-Aky2-EPpiMhHDDGQ/viewform

The second round of surveys can be tracked using the link:

https://docs.google.com/forms/d/e/1FAIpQLSd3dAQ6IYSywhhuEpviYGSPP46Zg-01T0HLAzEUH_2ruC403A/viewform?usp=sf_link.

## ACKNOWLEDGMENT

For their constructive feedback on the previous version of this paper, the authors thank Professor Abeer Elshater and the anonymous reviewers. The authors also thank colleagues in the course entitled "Urban Design: Theory and Practice" offered by Ain Shams University in Urban Design postgraduate studies (Fall 2021) for providing insights that helped develop this work.

## REFERENCES


Ahmed, T., Moeinaddini, M., Almoshaogeh, M., Jamal, A., Nawaz, I., & Alharbi, F. (2021). A New Pedestrian Crossing Level of Service (PCLOS) Method for Promoting Safe Pedestrian Crossing in Urban Areas. *International Journal of Environmental Research and Public Health*, *18*(16), 1–18. doi:10.3390/ijerph18168813 PMID:34444563

Anciaes, P. R., & Jones, P. (2018). Estimating preferences for different types of pedestrian crossing facilities. *Transportation Research Part F: Traffic Psychology and Behaviour*, *52*, 222–237. doi:10.1016/j.trf.2017.11.025

Appolloni, L., Giretti, A., Corazza, M., & D'alessandro, D. (2020). Walkable urban environments: An ergonomic approach of evaluation. *Sustainability*, *12*(20), 1–31. doi:10.3390u12208347

Arafa, A. E.-S., & Hirshon, J. M. (2019). Driving behavior and road traffic crashes among professional and nonprofessional drivers in South Egypt. *International Journal of Injury Control and Safety Promotion*, *26*(4), 372–378. doi:10.1080/17457300.2019.1638419 PMID:31282807

Ben-Bassat, T. (2019). Are ergonomically designed road signs more easily learned? *Applied Ergonomics*, *78*, 137–147. doi:10.1016/j.apergo.2019.02.009 PMID:31046945

Ben-Bassat, T., Shinar, D., Caird, J. K., Dewar, R. E., Lehtonen, E. S., Zakowska, L., ... Pronin, M. (2021). Ergonomic design improves cross-cultural road sign comprehension. *Transportation Research Part F: Traffic Psychology and Behaviour*, *78*, 267–279. doi:10.1016/j.trf.2021.01.015

Bhalla, K., & Gleason, K. (2020). Effects of vehicle safety design on road traffic deaths, injuries, and public health burden in the Latin American region: A modelling study. *The Lancet. Global Health*, *8*(6), 1–10. doi:10.1016/S2214-109X(20)30102-9 PMID:32446347

Chebanyuk, K., Prasolenko, O., Burko, D., Galkin, A., Lobashov, O., Shevchenko, A., Usami, D. S., & Persia, L. (2020). Pedestrians influence on the traffic flow parameters and road safety indicators at the pedestrian crossing. *Transportation Research Procedia*, *45*, 858–865. doi:10.1016/j.trpro.2020.02.083

Costa, A. P., Campos, F., & Villarouco, V. (2012). Overview of ergonomics built environment. *Work (Reading, Mass.)*, *41*, 4142–4148. doi:10.3233/WOR-2012-0710-4142 PMID:22317357

De Carvalho, R. (2012). ERGOPOLIS: An ergonomics approach applied to a city. *Work (Reading, Mass.)*, *41*, 6071–6078. doi:10.3233/WOR-2012-1063-6071

Dohrmann Consulting. (2014, July 5). *What is Ergonomics?* Retrieved January 2022, from Dohrmann Consulting: https://www.ergonomics.com.au/about-us/

Dul, J., Bruder, R., Buckle, P., Carayon, P., Falzon, P., Marras, W. S., Wilson, J. R., & van der Doelen, B. (2012). A strategy for human factors/ergonomics: Developing the discipline and profession. *Ergonomics*, *55*(4), 377–395. doi:10.1080/00140139.2012.661087 PMID:22332611

Eilouti, B. (2009). A digital incorporation of ergonomics into architectural design. *International Journal of Architectural Computing*, *7*(2), 235–253. doi:10.1260/147807709788921994

Eilouti, B. (2021). A framework for integrating ergonomics into architectural design. *Ergonomics in Design*, 1–9. doi:10.1177/10648046209983672

El Bardisy, N., Reuter, W., & Ismail, A. (2019). Social interaction dynamics exploration: The case Study of el sherouk city. *International Journal of Architectural and Environmental Engineering*, *13*(11), 658–667.

Eldar, R., & Fisher-Gewirtzman, D. (2019). Ergonomic design visualization mapping-developing an assistive model for design activities. *International Journal of Industrial Ergonomics*, *74*, 102859. Advance online publication. doi:10.1016/j.ergon.2019.102859

ElMoghazi, Y. (2019). *Road safety challenges in Egypt: A discussion of policy alternatives*. Retrieved 2022, from Theses and Dissertations: https://fount.aucegypt.edu/etds/790/

Elshater, A., Abusaada, H., Alfiky, A., El-Bardisy, N., Elmarakby, E., & Grant, S. (2022). Workers' Satisfaction vis-à-vis Environmental and Socio-Morphological Aspects for Sustainability and Decent Work. *Sustainability*, *14*(3), 1–25. doi:10.3390u14031699

Fayed, L., Elshater, A., & Rashed, R. (2020). Aspects of Regenerative Cities. In S. Kamel, H. Sabry, G. F. Hassan, M. Refat, A. Elshater, & A. S. Elrahman (Eds.), *R. Rashed, Architecture and urbanism: A smart outlook* (pp. 303–319). Springer. doi:10.1007/978-3-030-52584-2_22

Fusaro, G., & Kang, J. (2021). Participatory approach to draw ergonomic criteria for window design. *International Journal of Industrial Ergonomics*, *82*, 103098. Advance online publication. doi:10.1016/j.ergon.2021.103098

Hand, G., Weber, R., & Bluestone, N. (2017). Regenerative cities: Moving beyond sustainability–a Los Angeles case study. *Journal of Urban Design and Mental Health, 3*, 1–14.

Hassan, G., Rashed, R., & El Nagara, S. (2022). Regenerative urban heritage model: Scoping review of paradigms' progression. *Ain Shams Engineering Journal*, 1–8. doi:10.1016/j.asej.2021.101652

Healthcare, I., & Karwowski, W. (2006). *International Encyclopedia of Ergonomics and Human Factors - 3 Volume Set* (I. Healthcare & W. Karwowski, Eds.). Taylor & Francis Group. doi:10.1201/9780849375477

International Labour Organization. (2021). *Principles and guidelines for human factors/ ergonomics (HFE) design and management of work systems.* Retrieved 2022, from http://www.ilo.org/global/topics/safety-and-health-at-work/news/WCMS_826603/lang--en/index.htm

Jastrzebowski, W. (1857). *Outline of Ergonomics, or the science of Work Based Upon Truths Drawn from Science of Nature 1857.* Central Institute for Labour Protection.

Lin, Y., Sun, C. A., Lin, H., Perng, H., Chung, C., Lin, C., Lai, C.-H., Chien, W.-C., & Tsai, C. (2021). Increased risk of road traffic injuries in individuals with insomnia. *Journal of Transport & Health, 21*, 1–5. doi:10.1016/j.jth.2021.101030

Naci, H., Chisholm, D., & Baker, T. (2009). Distribution of road traffic deaths by road user group: a global comparison. *Injury Prevention, 15*(1), 55–59. doi:10.1136/ip.2008.018721

Occupational Safety and Health Administration. (1999). *Ergonomics Program.* U.S. DEPARTMENT OF LABOR.

Occupational Safety and Health Administration. (2022). *Occupational Safety and Health Administration.* Retrieved from https://www.dol.gov/

Pao, T.-H., & Kleiner, B. H. (2021). New developments concerning the occupational safety and health act. *Managerial Law, 43*(1/2), 138–146. doi:10.1108/03090550110770462

Puvanachandra, P., Hoe, C., El-Sayed, H. F., Saad, R., Al-Gasseer, N., Bakr, M., & Hyder, A. (2012). Road traffic injuries and data systems in Egypt: Addressing the challenges. *Traffic Injury Prevention, 13*(1), 44–56. doi:10.1080/15389588.2011.639417 PMID:22414128

Read, G., Salmon, P., Goode, N., & Lenné, M. (2018). A sociotechnical design toolkit for bridging the gap between systems-based analyses and system design. *Human Factors and Ergonomics in Manufacturing & Service Industries*, *28*(6), 327–341. doi:10.1002/hfm.20769

Saptaputra, S. K., Kurniawidjaja, L. M., Susilowati, I. H., & Pratomo, H. (2021). Ergonomic sofa design to support kangaroo mother care in Indonesia. *Journal of Neonatal Nursing*, *27*(6), 471–475. doi:10.1016/j.jnn.2021.06.013

Shoubi, M. V., Barough, A. S., & Rasoulijavaheri, A. (2013). Ergonomics principles and utilizing it as a remedy for probable work related injuries in construction projects. *International Journal of Advances in Engineering and Technology*, (Mar), 232–245.

Stevens, N. J., & Salmon, P. M. (2015). New knowledge for built environments: Exploring urban design from socio-technical system perspectives. in Artificial Intelligence and Lecture Note. In D. Harris (Ed.), *Engineering Psychology and Cognitive Ergonomics* (Vol. 9174). Springer. doi:10.1007/978-3-319-20373-7_19

Tayyari, F., & Smith, J. L. (1997). *Occupational Ergonomics: Principles and application*. Kluwer Academic Publishers.

The International Ergonomics Association (IEA). (2022). *History of The IEA*. Retrieved from IEA: https://iea.cc/history/

Wang, J., Han, S. H., & Li, X. (2021). 3D fuzzy ergonomic analysis for rapid workplace design and modification in construction. *Automation in Construction*, *123*, 1–14. doi:10.1016/j.autcon.2020.103521

Wilson, J. (2014). Fundamentals of systems ergonomics/human factors. *Applied Ergonomics*, *45*(1), 5–13. doi:10.1016/j.apergo 2013.03.021 PMID:23684119

World Health Organization. (2018). *Global status report on road safety 2018*. World Health Organization.

World Health Organization. (2021, June 21). *Road traffic injuries*. Retrieved 2022, from Newsrooms: https://www.who.int/news-room/fact-sheets/detail/road-traffic-injuries

World Health Organization. (2021, June 21). *Road traffic injuries*. Retrieved February 2022, from https://www.who.int/news-room/fact-sheets/detail/road-traffic-injuries

World Health Organization. (2022). *Estimated road traffic death rate (per 100 000 population)*. Retrieved 2022, from Rate: https://www.who.int/data/gho/indicator-metadata-registry/imr-details/198

Zafri, N. M., Rony, A. I., & Adri, N. (2019). Analysis of pedestrian crossing speed and waiting time at intersections in Dhaka. *Infrastructures*, *4*(3), 1–20. doi:10.3390/infrastructures4030039

## KEY TERMS AND DEFINITIONS

**Delphi Method:** Also known as the Delphi technique, is a structured communication methodology or process that was first established as an organized, interactive forecasting method based on a panel of experts. The technique, adapted for face-to-face meetings, is known as mini-Delphi or Estimate-Talk-Estimate.

**Ergonomics:** Is defined according to the International Ergonomics Association (IEA) as ergonomics (or human factors) as a scientific discipline concerned with the understanding of interactions among humans and other elements of a system and the profession that applies theory, principles, data and methods to design. IEA as ergonomics (or human factors) i a scientific discipline concerned with understanding interactions among humans and other elements of a system and the profession that applies theory, principles, data, and methods to design.

**Ergopolis:** In the context of urban planning, ergopolis refers to the idea of thinking and transforming cities from the point of understanding their inherent complexity and their interactions with people.

**Planning, Doing, Studying, and Acting (PDSA):** Is a method that outlines how changes can be planned and implemented in various areas. Following the prescribed four steps facilitates breaking down the process into smaller steps, evaluating the outcome, improving it, and testing again. In most cases, we go through some or all these steps when we implement changes in our lives without even thinking about it. However, writing them down may enable people to focus more effectively.

**Regenerative Cities:** A regenerative city is a type of urban development based on an environmentally beneficial, stimulating interaction with the natural systems from which it derives resources for its survival

# Chapter 9
# Mapping Heat Exposure of Pedestrian Density Around Metro Stations Using Artificial Intelligence:
## Ramses Square, Cairo, Egypt

**Shereen Wael**
*Ain Shams University, Egypt*

**Abeer Elshater**
https://orcid.org/0000-0002-5061-6861
*Ain Shams University, Egypt*

**Samy Afifi**
https://orcid.org/0000-0003-2315-9850
*Ain Shams University, Egypt*

## ABSTRACT

*Global research extensively discusses artificial intelligence (AI) and thermal comfort in urban studies. The current research design uses Goodvision Video Insights as an application of AI to analyse users' preferences for movement within three exits/entrances of the metro stations in Ramses Square, Cairo, Egypt. This study uses AI technology to track pedestrian movement based on heat exposure in metro station contexts. The primary findings identified the relationship between user experiences and urban composition. These results also revealed that the urban morphology of the metro station context impresses users' thermal comfort and movement preferences enough to track their needs through commutes. This study could not address all the aspects that influence users' experiences in the case study. Future studies provide light on the impact of varying temperature conditions on comfort levels and, as a result, human behaviour in station areas.*

DOI: 10.4018/978-1-6684-2462-9.ch009

## INTRODUCTION

A massive literature group tackles the challenges of climate change and smart technologies in urban research. A group of studies looks at the effects of urban heat islands (UHIs) on cities and how this phenomenon affects user perceptions (Fang, et al., 2021; Roshan, Oji, & Fdez-Arróyabe, 2022; Hernandez & Monzon, 2016; Hutton, 2013). Numerous studies explore the urban heat island effect, which arises when cities replace the natural land cover with high pavement, buildings, and other heat-absorbing and heat-retaining surfaces (Elmarakby, Khalifa, Elshater, & Afifi, 2021; Elbardisy, Salheen, & Fahmy, 2021). This impact results in a rise in energy expenses (for example, air conditioning), air pollution levels, and heat-related unhealthy environments. Scholars address artificial intelligence (AI) to get the benefits of digital research methods. The literature describes AI as an influential technology with increasing applications in various fields. AI is the technology that enables machines to do cognitive functions that were always done by humans (Yigitcanlar, et al., 2020). While cities are now increasingly digitised, technology's role in urban fabrics has been widely covered (Hashem, et al., 2016). Literature acknowledges smart cities by using AI applications as an efficient tool to observe and model the creation of sustainable urban environments (Abusaada & Elshater, 2021a; Gessa & Sancha, 2020; Yigitcanlar, et al., 2018).

In this chapter, among various actions for achieving sustainable urban environments, we adapt the metro as a sustainable mode of public transportation (Falvo, Lamedica, Bartoni, & Maranzano, 2011; Khosravi, Haghshenas, & Saleh, 2020). The previous literature discussed how urban designers and planners play an essential role in reorienting people's daily travel decisions (Stojanovski, 2020; Middleton, 2011; Salonen & Toivonen, 2013). Providing a comfortable urban environment around stations may encourage people to use public transit (Rijsman, van Ort, Hoogendoorn, Molin, & Teijl, 2019). Capturing the behaviour of pedestrians in various situations is essential in urban planning, land use, marketing, or traffic operations (Antonini, Bierlaire, & Weber, 2005; Boarnet & Crane, 2001). Previous studies relied on several methods to provide qualitative and quantitative analysis of human behaviour in urban spaces (Rahaman, Ojmori, & Harata, 2005; Sarkar, 2003; Zakaria & Ujang, 2015).

This chapter tackles the challenges of metro transit as a primary transportation mode in Cairo, with more than three million people using the current three metro lines daily (Cairo Governorate Metro, 2021). This massive crowd of people might face problems in their daily commutes because of external factors like heat exposure. Metro commuters always need to use other public transportation methods or walk to reach their destinations. There are various pedestrian scenarios around metro stations.

Pedestrians on Egyptian metro lines deal with different conditions throughout their door-to-door journey, from origin to destination.

Several studies have deliberated on the effects of outdoor thermal conditions on pedestrian behaviour in various shapes and urban spaces (Fahed, Ginestet, & Adolphe, 2020; Hendel, Azos-Diaz, & Tremeac, 2017; Sun, Zacharias, Ma, & Oreskovic, 2016). However, there is a lack of studies that use various AI applications to detect and model pedestrian behaviour and link it to changing thermal conditions. The gap in the literature is in providing quantitative analysis and mapping pedestrian movement patterns that can reflect their experience in a more precise way (Elias, 2011; Lesani & Miranda-Moreno, 2018). This study helped investigate the effect of the emergence of UHI effects and their physical elements on users' experiences and comfort in the study area.

This chapter investigates the relationship between outdoor thermal conditions and pedestrian movement behaviour and preferences by using AI applications to detect and map users' experiences in the station's context. This study had two main aims: first, to investigate the capability of AI applications to detect and analyse human behaviour in urban spaces. The second aim was to study the UHI effect on pedestrian movement behaviour in these areas. The researchers worked to fill the literature gap on using technology in evaluating users' experiences in different urban environments.

This study focuses on studying pedestrian behaviour at three station exits in Ramses, Cairo. It investigates the relationship between outdoor thermal conditions and users' experiences based on the study of the urban context. The results showed the role of AI technology in mapping people behaviour in the stations' context and its link with thermal conditions. The pedestrian movement study depended on one of the ascending AI applications named "Goodvision Video Insights." It is a complex software platform that provides multiple features for collecting and analysing traffic data. In addition, the study relied on thermal imaging techniques to map thermal conditions in the study area. The study results could guide the development plans for Egypt's existing and future metro stations.

## BACKGROUND

This section defines the concepts of UHI as one of the factors affecting users' experiences in public spaces, followed by the role of AI in mapping human behaviour in public places. UHI effect is a kind of heat accumulation phenomenon within the urban area due to urban construction and various human activities (Yang, Qian, Song, & Zheng, 2016). Previous literature has tackled the UHI effect at multiple scales: microscale (streets), local scale (neighbourhood, district), or macroscale (city) (Elmarakby, Khalifa, Elshater, & Afifi, 2021; Faragallah & Ragheb, 2022). Several studies analysed urban activities that increase the UHI caused by the increase in heat emission from the ground surface of dense urban areas. Most of the heat radiation is reflected from paving materials, e.g., asphalt and concrete covers, that generate high temperatures, in addition to the absence of vegetation in urban spaces (Long L. & Zhang, 2020). Rising air pollution levels and the variation in surface temperatures seriously affect human comfort (Fang, et al., 2021). Consequently, heat stress is one of the factors that makes pedestrians avoid walking, riding, and reducing short-distance travel and physical activities (Irmak, Yilmaz, & Dursun, 2017; Salonen & Toivonen, 2013).

Human thermal comfort expresses their satisfaction with the thermal environment. It is mainly influenced by the energy exchanges between the human body and the surrounding context (Hiemstra, Saaroni, & Amorim, 2017). Outdoor thermal comfort is critical in urban studies (Melnikov, Krzhizhanovskaya, & Sloot, 2017). Continued exposure to stressful heat is not only an uncomfortable experience but also threatens human health. There are several ways of assessing thermal comfort. These ways depend on a range of physical characteristics of the environment and the human comfort experience. Air temperature is the most common parameter in studies of UHIs. In addition, it is also the most easily measured determinant of thermal comfort. The radiant environment has a significant impact on thermal comfort. The variation of radiant field is highly dependent on context and variation of the surface temperature of a few degrees that may lead to discomfort. Several factors, e.g., direct solar radiation, airspeed and humidity (Watkins, Palmer, & Kolokotroi, 2007).

Today's countries have active transportation initiatives to reduce urban heat islands (UHI) by reducing automobile use in cities and promoting cycling and walking. There is an international trend to maximise dependence on sustainable modes of transportation to mitigate environmental problems (Cottrill & Derrible, 2015; Dai, Sun, & Xu, 2018). Metro station has two basic identities based on its function. It is a node as it is the point of access to Metro cars and other transportation methods if they exist. It is also an important landmark in its district. At the same time, it is a place of convergence. It reflects human activities and their flow across the city. Scanning previous studies, stations are nodes for transportation lines and nodes for human activities. This makes stations include spatial interactions among urban places, economic activities and people (Bertolini, 2008). The station as a node has unique functions in its district life. It creates a different affective atmosphere. While the metro station is a place, it is a physical environment that has a mutual effect on physical surroundings.

The thermal characteristics of outdoor urban spaces and the street networks connecting them are vital to assessing the livability and sustainability of cities. Urban design is strongly dependent on climate interactions, which improve or moderate impacts on human thermal comfort. Vanos (2015) suggested that poorly designed public spaces in districts result in stressed urban heat, thus increasing the outdoor temperature and seriously impacting outdoor human activities. Shaded pedestrian and cyclist paths and more relaxed environments contribute to encouraging walking, cycling, or even transferring between different public transportation means (Alattar & Furlan, 2017). Elnabawi & Hamza (2020) highlight the importance of built material and design on meteorological conditions and thermal perceptions. The use of appropriate urban geometry and vegetation seems to be very promising in mitigating the adverse UHIs effect and providing a better pedestrian thermal comfort (Jamei, Rajagopalan, Seyedmahmoudian, & Jamei, 2015). Bourbia & Awbi (2004) showed that by controlling the street architecture, it is possible to prevent high temperatures in urban canyons and therefore has an effect on a local scale rather than city scale. Based on the above, the design of stations' context is an essential factor that affects commuters' thermal comfort. There is strong evidence available that demonstrates that improvements to the door-to-door (from origin to destination) journey can increase usage of public transport (Tovey, Woodcock, & Osmond, 2017). Consequently, the stations' surrounding urban form could influence people preferences of the transportation means use.

The development of AI applications contributes to comprehensively studying and assessing people's movement patterns, behaviour, and preferences in the context of the stations (Kim, 2020). Most pedestrian detection studies were related to computer vision techniques for detecting people in a series of images, whether in live streams, recorded videos or captured photos (Basbasa, Campisi, & Canale, 2020; Navarro, Fernández, Borraz, & Alonso, 2017). Computer vision identifies and detects objects and structures within images (Kajabad & Ivanov, 2019). Data scientists have trained their models to see the targeted things. A computer vision pipeline is a series of main steps its applications go to process the input data (Figure 1). Its technique provides accurate data for the detected objects: count, speed, density, patterns, etc. The output is quantitative data that describes the situation of a detected object through numbers, statistics and heatmaps. When people are the target, computer vision techniques contribute to mapping people behaviour in various areas. Consequently, urban designers, planners and researchers would be able to investigate people's needs and preferences in the context of the stations.

*Figure 1. Computer vision pipeline*

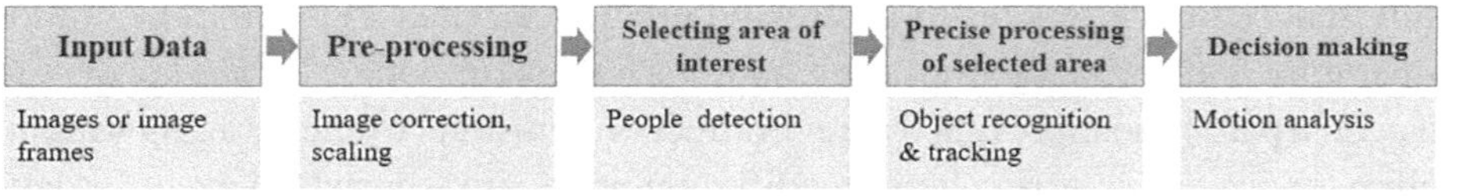


# METHOD

## Study Area

This research investigated users' experiences at three metro station exits/entrances in Ramses, Cairo. The study focused on mapping these experiences in a vital area of a metro station. We selected Orabi Metro Station in Cairo downtown, leading to Ramses Street, surrounded by critical establishments. Most of the surrounding buildings are administrative, public service, and commercial in this selected site. Orabi Station almost always has daily and occasional commuters due to the surrounding area. This research studied such a vital area with various commuters and activities. The study area is around 18,000 square meters, including three-station exits. The surrounding buildings are at the proper height; that helped record videos from the top (Figure 2).

## Data Processing

The primary stage was to precisely observe the station's area and map the whole situation of commuters' behaviour. The researchers determined the rush hours through the initial observation of vehicular and pedestrian movement daily. Rush hours are the target period to record and assess pedestrian movement and preferences. The study prepared a digitised map including urban configuration physical elements. Data was collected by direct observation through the site visit on March 14, 2021. ArcGIS pro tools helped develop visibility analysis and shadow estimation of the station's surrounding buildings, trees, and other street furniture elements. It built a database for buildings' uses, heights, street networks, sidewalks, plot patterns and different landscape architecture locations to assess the surrounding urban environment.

The researchers of the present work mapped the thermal environment using a thermal imaging technique. Thermal imaging is used to examine UHIs using a high-resolution thermal infrared camera. The thermal camera receives radiation from targeted objects and their surroundings (Elmarakby, Khalifa, Elshater, & Afifi, 2021). The researchers used the FLIR T540 Thermal Camera (model: 79302-0201) to capture photos from various points in the study area to address the entire thermal

*Figure 2. Location of the study area*

states of objects. The images were captured from 1:00 to 2:00 pm on December 16, 2021, the selected period of rush hours. The thermal images were processed using the FLIR Tools software. It worked on mapping temperature variation in the study area. The previous process produced photos with heat maps of colour representing the temperature variation in the station's areas. By observing these photos, the researchers would investigate the impact of urban configuration on the thermal conditions in different regions.

The current study relied on 'GoodVision Insights' software. GoodVision Insights is an application that relies on computer vision technology to detect both people and vehicular movement. It provides deep traffic data analytics through live streams or recorded videos. Input data was recorded using an ordinary camera at three entrances from the tops of surrounding buildings in March 2021. The researchers captured

*Figure 3. Camera's location with videos cadre*

videos of the entrance area in three cadres within a radius of around 40 meters. Four videos per day were recorded at rush hours on specific days, avoiding non-working ones for one month to study people's movement at the highest densities (Figure 3).

The primary stage held by the application was to correct and scale the series of images of the recorded videos. The videos' processing stage extracted traffic data to define the whole situation. The researchers divided each cadre into cells of equal areas by drawing cross grids. The next step was to generate numeric data for each cell. The areas of interest had to be defined for the application to extract their data. Here, the study's central area of interest was people who needed to detect, count, and map their movement behaviour. This application relies on people passing by cells specified by the researchers. The researchers had to divide the study area into cells into the same area of the GIS map. Then, the researchers entered the numeric output data into the GIS database to form accurate heatmaps with numeric values. This step helped link this type of data for pedestrian behaviour to the study of urban configuration and compare the density values in both areas.

Researchers in the urban field need precise detection of people moving in the stations' area to perceive human movement behaviour and preferences. Video processing produced raster images of heatmaps with colour ranges for speed and held up areas. It also provided illustrations of movement paths in the study area. These maps were projected onto the previously prepared map to produce accurate heatmaps that the researchers could compare various sites. The yielded illustrations helped to observe pedestrians' preferred paths. This helped detect pedestrian movement, orientation, and which areas they preferred to stop or pass through. Motion detection also helped investigate commuters' leading destinations and the routes used to reach them.

Automatic pedestrian detection and counting can provide reliable data collection and descriptive statistics for the entire urban context (Abusaada & Elshater, 2020; Kim, 2020; Sohn, 2013). This type of data is essential to investigate pedestrian behaviour and experience based on the previous literature. This application could clearly and precisely determine pedestrian density and flow rate using computer vision technology. The output data helped the current study prepare descriptive statistics for people density and flow rate on the different days for the station's entrances area. In addition, it showed the places where people prefer to stop or rest.

## RESULTS AND DISCUSSION

Results from the thermal camera output found that low-temperature areas were mainly attached to the surrounding buildings. The temperature of the sidewalks of interlocking tiles was consistently lower than that of asphalt. In addition, concrete curbs were the most element gaining heat in all areas. Thermal photos showed that buildings with their footprint, height and orientation contributed to lowering the temperature in the station's area by providing shade and wind flow allowance. Buildings in exits 2 and 3 have played a significant role in shading the adjacent sidewalks and lowering their temperature. This was because of the ratio between the buildings' height and the width of the sidewalk in addition to their location (Figure 4, 5).

By comparing the two areas of the station's exits, the researchers found that trees also played the leading role in lowering the temperature in exit 2, 3 opposing exit 1 areas. On the other side, vendors and kiosks were always using the tree-

*Figure 4. Thermal images for exit 1 area*

shaded areas as a place for their activity. Exits' structures also were gaining more heat than other adjacent elements. People density heat maps produced from videos showed that higher densities were always beside the surrounding buildings where the temperature was lower than other areas most of the time. Pedestrian density in exit 2, 3 area was more distributed along the width of the sidewalk than exit 1 area. Density distribution was strongly related to the hierarchical variation of temperature in both areas (Figure 6).

As for pedestrian movement behaviour, the researchers found that people preferred to use paths with lower temperatures most of the time. Furthermore, many pedestrians were forced to walk or stop in warmer areas to reach their destinations or wait for other transportation means (Figure 6, Figure 8). Pedestrian speed change heatmap showed that pedestrian speed was related to the density and flow rates. Speed heatmaps also showed that people were walking faster in warmer areas than in more excellent areas (Figure 9). People walk faster at temperatures outside their comfortable range to escape the uncomfortable area as soon as possible (Melnikov, Krzhizhanovskaya, & Sloot, 2017). This could reflect the effect of thermal comfort in various ranges on pedestrian behaviour and movement patterns. This effect was

*Figure 5. Thermal images for exit 2, 3 area*

more apparent by relating the thermal photos to the pedestrian preferred and most used paths.

The people detection process also produced hold-up heatmaps that showed the distribution of areas according to the people's time. The researchers found that people mostly had no chance to stop or relax in the exit's area. A few people rested on the exit's structure for seconds (Figure 10). This situation could be related to the higher temperatures around the exits and poor street furniture, and lack of shaded and open spaces. The presence of stopping places was critical in studying the comfort level around the metro station.

The findings from the impacts of contextual influences on pedestrian movement in city streets corroborate other studies that propose taking these aspects into account when designing more walkable spaces (Abusaada & Elshater, 2021b; Abusaada & Elshater, 2019). It is worth pointing out that some authors have used this methodology for UHI effects on pedestrian movement and the whole commute experiences (Elmarakby, Khalifa, Elshater, & Afifi, 2021; Fahed, Ginestet, & Adolphe, 2020). Several authors have explored the effects of contextual design as a potential option for better walkable places (Tarek, Hassan, Elshater, & Elfayoumy, 2021; Abusaada &

*Figure 6. Density heatmaps in station's exits area*

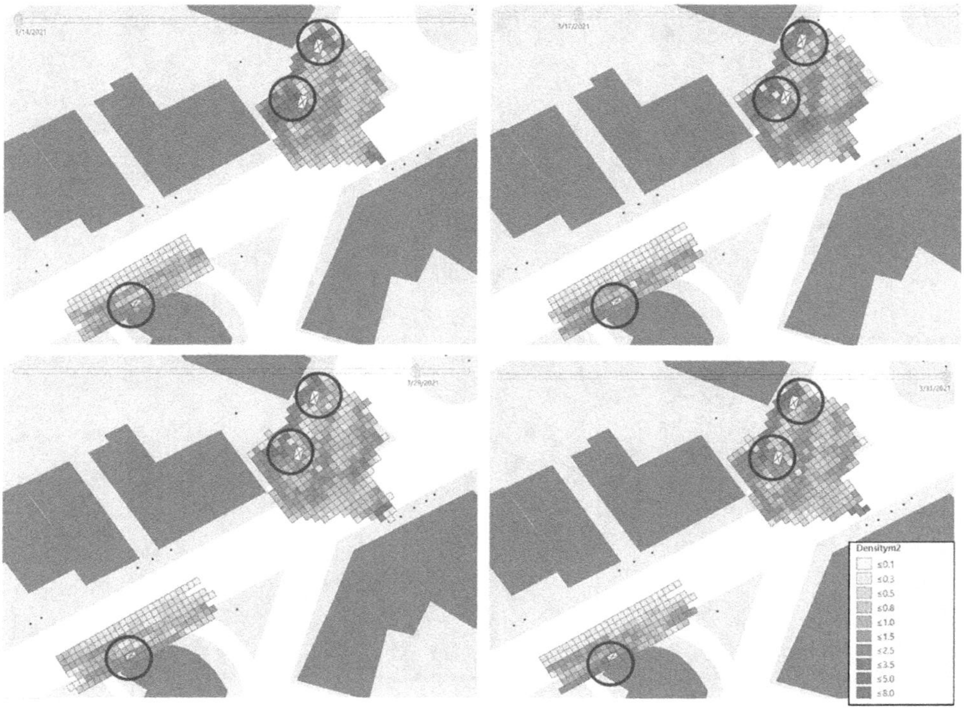


Elshater, 2021c; Abusaada, 2020). Our results have been shown to work particularly well to improve users' satisfaction in public places.

This research has faced several limitations. Security issues were one of the most robust limitations as there were obstacles to recording videos, fixing cameras, or using thermal cameras in some areas. However, this research recorded limited videos on different days for the Orabi Metro station area. Consequently, this work could not cover all factors that affect users' experience in the study area. A final limitation was the lack of technical support that may help relate all results automatically instead of the direct observation.

This study investigated users' experiences using AI technology in Cairo's metro station's context. It showed that thermal conditions other elements that form the urban context shape users' experience. This research provides a method of using AI technology to study and develop urban areas.

The contribution here is to shed light on the planning and design process of the station context and their level of services. This research proved that the design and planning processes should use technology to track and evaluate people's behaviour in the stations' area. The study investigated the effect of urban configuration with

*Figure 7. Pedestrian movement paths in exit 1 areas*

*Figure 8. Pedestrian movement paths in exit 2, 3 areas*

*Figure 9. Pedestrian speed change map*

*Figure 10. Pedestrian hold-up heatmaps*

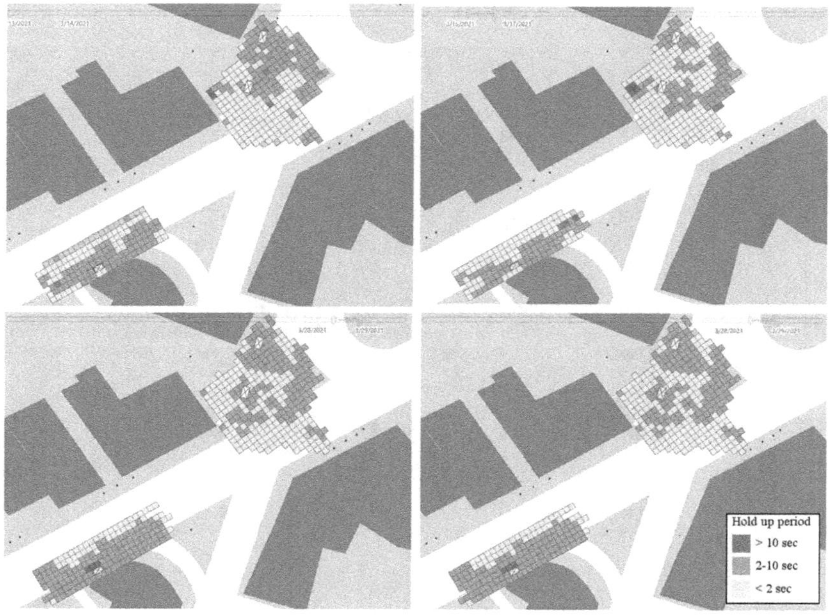

its physical element on pedestrian thermal comfort and consequently their choices. The contribution made here also is a widespread application of AI technology use in controlling the development plans and decisions in stations' context. The added value of this chapter is in providing an applicable method to study people behaviour in urban areas through both quantitative and qualitative data.

## CONCLUSION

The literature documenting human behaviour and pedestrian movement around metro stations is challenging. This research discusses artificial intelligence (AI) and thermal comfort in urban studies, focusing on Ramses Square in Cairo. Using AI technology, this study tracked pedestrian movement based on heat exposure in metro station contexts. This study mapped pedestrian movement and their preferences for movement—as attachments to place—in the metro station contexts that best fit their needs and provide a convenient experience. The current research design uses AI to analyse users' preferences for movement within three exits/entrances of the metro stations in Ramses Square. The results shed light on the role of AI technology in mapping human behaviour in a physical context. The main primary findings identified the inherent relationship between user experiences and urban composition. These results also revealed that the urban morphology of the metro station context impresses users' thermal comfort and movement preferences enough to track their needs through commutes. This study opens future research directions with the potential to develop a qualitative and quantitative analysis of people's experiences around metro stations.

The results confirmed that AI with computer vision applications could play a significant role in tracking and studying human behaviour in a more precise way. Computer vision applications provide complete photographs of users' experiences through numeric values and descriptive statistics. Heatmaps produced from this type of application help urban researchers to recognise the users' experience and relate these maps to the existing conditions. Consequently, using technology and innovative techniques can facilitate design development plans for new and existing stations. In addition, this chapter confirmed the role of urban designers in shaping users' experiences in the station's area.

The results investigated the effect of urban configuration and its physical elements on pedestrian thermal comfort and overall experience. Different features, e.g., building structures, paving materials, and vegetation, affect the thermal conditions of the station's context. Pedestrian pathway detection showed a strong relationship between thermal requirements and people's preferences. People density, flow, and speed, which are the leading indicators of the level of service of urban areas, were strongly affected by temperature variations in the different regions. The researchers could not perceive this effect through the previous literature. People seek to gain the highest level of thermal comfort by preferring cooler and more shaded areas. It was clear through relating thermal photos of the station's site with people density and movement maps. Thermal conditions also seriously affected pedestrian speed changes and their comfort levels.

Future research may address the existing work limits by investigating diverse urban settings in additional station regions. Technical professionals should assist future research teams in connecting the findings. In terms of security, future studies may work with the government to install cameras in various regions to monitor people's behaviour on different days. This proposed region of investigation would aid in determining movement patterns throughout the day and even months. As a result, further study might help understand the impact of different temperature conditions on comfort levels and human behaviour in station areas. These findings would aid in formulating criteria for station area design and development. Future study designs may incorporate other qualitative data collection methods, such as interviews or questionnaires, to compare with the findings of human behaviour monitoring. This chapter described a technique for mapping users' experiences in the metro station using AI technology to help the development and design processes may be a beneficial tool for understanding human behaviour and preferences.

## REFERENCES

Abusaada, H. (2020). Strengthening the affectivity of atmospheres in urban environments: The toolkit of multi-sensory experience. *Archnet-IJAR*, *14*(3), 379–392. doi:10.1108/ARCH-03-2020-0039

Abusaada, H., & Elshater, A. (2019). Effects of urban atmospheres on changing attitudes of crowded public places: An action plan. *International Journal of Community Well-Being*, *3*(2), 109–159. doi:10.100742413-019-00042-w

Abusaada, H., & Elshater, A. (2020). Urban design assessment tools: A model for exploring atmospheres and situations. *Proceedings of the Institution of Civil Engineers. Urban Design and Planning*, *173*(6), 238–255. doi:10.1680/jurdp.20.00025

Abusaada, H., & Elshater, A. (2021a). Competitiveness, distinctiveness and singularity in urban design: A systematic review and framework for smart cities. *Sustainable Cities and Society*, *68*, 1–12. doi:10.1016/j.scs.2021.102782

Abusaada, H., & Elshater, A. (2021b). Effect of people on placemaking and affective atmospheres in city streets. *Ain Shams Engineering Journal*, *12*(3), 3389–3403. doi:10.1016/j.asej.2021.04.019

Abusaada, H., & Elshater, A. (2021c). Improving visitor satisfaction in Egypt's Heliopolis historical district. *Journal of Engineering and Applied Sciences (Asian Research Publishing Network)*, *68*(1), 1–22. doi:10.118644147-021-00022-y

Alattar, D. A., & Furlan, R. (2017). Urban regeneration in Qatar: A comprehensive planning strategy for the transport oriented development of Al-Waab. *Journal of Urban Regeneration and Renewal*, *11*(2), 168–193.

Antonini, G., Bierlaire, M., & Weber, M. (2005). Discrete choice models of pedestrian walking behavior. *Transportation Research Part B: Methodological*, *40*(8), 667–687. doi:10.1016/j.trb.2005.09.006

Basbasa, S., Campisi, T., & Canale, A. (2020). Pedestrian level of service assessment in an area close to an under-construction metro line in Thessaloniki, Greece. *Transportation Research Procedia*, *45*, 95–102. doi:10.1016/j.trpro.2020.02.067

Bertolini, L. (2008). Station areas as nodes and places in urban networks: An analytical tool and alternative development strategies. In Railway Development (pp. 35–57). Physica-Verlag HD. doi:10.1007/978-3-7908-1972-4_3

Boarnet, M., & Crane, R. (2001). The influence of land use on travel behavior: Specification and estimation strategies. *Transportation Research Part A, Policy and Practice*, *35*(9), 823–845. doi:10.1016/S0965-8564(00)00019-7

Bourbia, F., & Awbi, H. (2004). Building cluster and shading in urban canyon for hot dry climate: Part 1: Air and surface temperature measurements. *Renewable Energy*, *29*(2), 249–262. doi:10.1016/S0960-1481(03)00170-8

Cairo Governorate Metro. (2021). Retrieved 2020, from Cairo Metro Website: https://cairometro.gov.eg/en/about/1

Cottrill, C. D., & Derrible, S. (2015). Leveraging big data for the development of transport sustainability indicators. *Journal of Urban Technology*, *22*(1), 45–64. doi:10.1080/10630732.2014.942094

Dai, X., Sun, L., & Xu, Y. (2018). Short-term origin-destination based metro flow prediction. *Journal of Advanced Transportation*, *2018*, 1–18. doi:10.1155/2018/5942763

Elbardisy, W. M., Salheen, M. A., & Fahmy, M. (2021). Solar irradiance reduction using optimized green infrastructure in arid Hot regions: A case study in El-Nozha District, Cairo, Egypt. *Sustainability*, *13*(17), 9617. Advance online publication. doi:10.3390u13179617

Elias, A. (2011). Automobile-oriented or complete street? Pedestrian and bicycle level of service in the new multimodal paradigm. *Transportation Research Record: Journal of the Transportation Research Board*, *2257*(1), 80–86. doi:10.3141/2257-09

Elmarakby, E., Khalifa, M., Elshater, A., & Afifi, S. (2021). Tailored methods for mapping urban heat islands in Greater Cairo Region. *Ain Shams Engineering Journal*, *13*(2), 1–12. doi:10.1016/j.asej.2021.06.030

Elnabawi, M. H., & Hamza, N. (2020). Behavioural perspectives of outdoor thermal comfort in urban areas: A critical review. *Atmosphere*, *11*(1), 1–23. doi:10.3390/atmos11010051 PMID:32461808

Fahed, J., Ginestet, S., & Adolphe, L. (2020). Impact of urban heat island mitigation measures on microclimate and pedestrian comfort in a dense urban district of Lebanon. *Sustainable Cities and Society*, *61*, 1–14. doi:10.1016/j.scs.2020.102375

Falvo, M. C., Lamedica, R., Bartoni, R., & Maranzano, G. (2011). Energy management in metro-transit systems: An innovative proposal toward. *Electric Power Systems Research*, *81*(12), 2127–2138. doi:10.1016/j.epsr.2011.08.004

Fang, Z., Tang, T., Zheng, Z., Zhou, X., Liu, W., & Zhang, Y. (2021). Thermal responses of workers during summer: An outdoor investigation of construction sites in South China. *Sustainable Cities and Society*, *66*, 1–14. doi:10.1016/j.scs.2020.102705

Faragallah, R. N., & Ragheb, R. (2022). Evaluation of thermal comfort and urban heat island through cool paving materials using ENVI-Met. *Ain Shams Engineering Journal*, *13*(3), 1–13. doi:10.1016/j.asej.2021.10.004

Gessa, A., & Sancha, P. (2020). Environmental open data in urban platforms: An approach to the big data life cycle. *Journal of Urban Technology*, *27*(1), 27–45. do i:10.1080/10630732.2019.1656934

Hashem, I. A., Chang, V., Anuar, N. B., Adewole, K., Yaqoob, I., Gani, A., ... Chirmona, H. (2016). The role of big data in smart city. *International Journal of Information Management*, *35*(5), 748–758. doi:10.1016/j.ijinfomgt.2016.05.002

Hendel, M., Azos-Diaz, K., & Tremeac, B. (2017). Behavioral adaptation to heat-related health risks in cities. *Energy and Building*, *152*, 823–829. doi:10.1016/j.enbuild.2016.11.063

Hernandez, S., & Monzon, A. (2016). Key factors for defining an efficient urban transport interchange: Users' perception. *Cities (London, England)*, *50*, 158–167. doi:10.1016/j.cities.2015.09.009

Hiemstra, J. A., Saaroni, H., & Amorim, J. H. (2017). The Urban Heat Island: Thermal Comfort and the Role of Urban Greening. In D. Pearlmutter & C. Calfapietra (Eds.), *The Urban Forest* (pp. 7–19). Springer. doi:10.1007/978-3-319-50280-9_2

Hutton, B. (2013). *Planning Sustainable Transport*. Routledge. doi:10.4324/9780203436066

Irmak, M. A., Yilmaz, S., & Dursun, D. (2017). Effect of different pavements on human thermal comfort conditions. *Atmosfera*, *30*(4), 1–12. doi:10.20937/ATM.2017.30.04.06

Jamei, E., Rajagopalan, P., Seyedmahmoudian, M., & Jamei, Y. (2016). Review on the impact of urban geometry and pedestrian level greening on outdoor thermal comfort. *Renewable & Sustainable Energy Reviews*, *54*, 1002–1017. doi:10.1016/j.rser.2015.10.104

Kajabad, E. N., & Ivanov, S. V. (2019). People Detection and Finding Attractive Areas by the use of Movement Detection Analysis and Deep Learning Approach. *Procedia Computer Science*, *156*, 327–337. doi:10.1016/j.procs.2019.08.209

Khosravi, S., Haghshenas, H., & Saleh, V. (2020). Macro-scale evaluation of urban transportation demand management policies in CBD by using system dynamics case study: Isfahan CBD. *Ransportation Research Procedia*, 2671–2689. doi:10.1016/j.trpro.2020.08.246

Kim, D. (2020). Pedestrian and Bicycle Volume Data Collection Using Drone Technology. *Journal of Urban Technology*, *27*(2), 45–60. doi:10.1080/10630732.2020.1715158

Lesani, A., & Miranda-Moreno, L. (2018). Development and testing of a real-time WiFi-Bluetooth system for pedestrian network monitoring, classification, and data extrapolation. *IEEE Transactions on Intelligent Transportation Systems*, *20*(4), 1484–1496. doi:10.1109/TITS.2018.2854895

Long, L. Y. Z., & Zhang, J. (2020). Spatial and dynamic perspectives on surface urban heat island and their relationships with vegetation activity in Beijing, China, based on Moderate Resolution Imaging Spectroradiometer data. *International Journal of Remote Sensing*, *41*(3), 882–896. doi:10.1080/01431161.2019.1650985

Melnikov, V., Krzhizhanovskaya, V. V., & Sloot, P. M. (2017). Models of pedestrian adaptive behaviour in hot outdoor public spaces. *Procedia Computer Science*, *108*, 185–194. doi:10.1016/j.procs.2017.05.006

Middleton, J. (2011). "I'm on autopilot, I just follow the route": Exploring the habits, routines, and decision-making practices of everyday urban mobilitie. *Environment & Planning*, *43*(12), 2857–2877. doi:10.1068/a43600

Navarro, P. J., Fernández, C., Borraz, R., & Alonso, D. (2017). A machine learning approach to pedestrian detection for autonomous vehicles using high-definition 3D range data. *Sensors (Basel)*, *17*(1), 1–20. doi:10.339017010018 PMID:28025565

Rahaman, K. R., Ojmori, N., & Harata, N. (2005). Evaluation of the roadside walkway environment of Dhaka City. *Proceedings of the Eastern Asia Society for Transportation Studies*.

Rijsman, L., van Ort, N., Hoogendoorn, S., Molin, E., & Teijl, T. (2019). Walking and bicycle catchment areas of tram stops: Factors and insights. In *6th International Conference on Models and Technologies for Intelligent Transportation Systems* (pp. 1–5). IEEE Xpolre. 10.1109/MTITS.2019.8883361

Roshan, G., Oji, R., & Fdez-Arróyabe, P. (2022). Future impact of cloudiness and wind changes on thermal comfort conditions in the city of Tabriz (Iran) in the frame of climate change. *Sustainable Cities and Society*, *77*, 1–15. doi:10.1016/j.scs.2021.103575

Salonen, M., & Toivonen, T. (2013). Modelling travel time in urban networks: Comparable measures for private car and public transport. *Journal of Transport Geography*, *31*, 1–11. doi:10.1016/j.jtrangeo.2013.06.011

Sarkar, S. (2003). Qualitative evaluation of comfort needs in urban walkways in major activity centers. *Transportation Quarterly*, *57*, 39–59.

Sohn, K. (2013). Feature mapping the Seoul Metro station areas based on a self-organizing map. *Journal of Urban Technology*, *20*(4), 23–42. doi:10.1080/10630732.2013.855514

Stojanovski, T. (2020). Urban design and public transportation–public spaces, visual proximity and Transit-Oriented Development (TOD). *Journal of Urban Design*, *25*(1), 134–154. doi:10.1080/13574809.2019.1592665

Sun, G., Zacharias, J., Ma, B., & Oreskovic, N. (2016). How do metro stations integrate with walking environments? Results from walking access within three types of built environment in Beijing. *Cities (London, England)*, *56*, 91–98. doi:10.1016/j.cities.2016.03.001

Tarek, M., Hassan, G. F., Elshater, A., & Elfayoumy, M. (2021). Investigating built environment indicators to develop a local walkability index. *Journal of Contemporary Urban Affairs*, *5*(2), 235–251. doi:10.25034/ijcua.2021.v5n2-7

Tovey, M., Woodcock, A., & Osmond, J. (2017). *Designing Mobility and Transport Services*. Taylor and Francis.

Vanos, K. J. (2015). Children's health and vulnerability in outdoor microclimates: A comprehensive review. *Environment International*, *76*, 1–15. doi:10.1016/j.envint.2014.11.016 PMID:25497108

Watkins, R., Palmer, J., & Kolokotroi, M. (2007). Increased temperature and intensification of the urban heat island: Implications for human comfort and urban design. *Built Environment*, *33*(1), 85–96. doi:10.2148/benv.33.1.85

Yang, L., Qian, F., Song, D.-X., & Zheng, K.-J. (2016). Research on urban heat-island effect. *Procedia Engineering*, *169*, 11–18. doi:10 1016/j.proeng.2016.10.002

Yigitcanlar, T., Kamruzzaman, M., Buys, L., Ioppolo, G., Sabatini-Marques, J., da Costa, E. M., & Yun, J. H. J. (2018). Understanding 'smart cities': Intertwining development drivers with desired outcomes in a multidimensional framework. *Cities (London, England)*, *81*, 145–160. doi:10.1016/j.cities.2018.04.003

Yigitcanlar, T., Kankanamge, N., Regona, M., Maldonado, A. R., Rowan, B., Ryu, A., ... Li, R. Y. (2020). Artificial intelligence technologies and related urban planning and development concepts: How are they perceived and utilized in Australia? *Journal of Open Innovation*, *6*(4), 1–21. doi:10.3390/joitmc6040187

Zakaria, J., & Ujang, N. (2015). Comfort of walking in the city center of Kuala Lumpur. *Procedia: Social and Behavioral Sciences*, *170*, 642–652. doi:10.1016/j.sbspro.2015.01.066

# Chapter 10
# Wind Effect on Pedestrian Activities and Motion Patterns:
## Istanbul Bilgi University Central Campus, Istanbul, Turkey

**Abdalrahman T. Y. Alashi**
https://orcid.org/0000-0003-2636-3013
*Istanbul Technical University, Turkey*

**Turgay Kerem Koramaz**
*Istanbul Technical University, Turkey*

**Dilek Yildiz Ozkan**
*Istanbul Technical University, Turkey*

**Özhan Ertekin**
*Istanbul Technical University, Turkey*

## ABSTRACT

*The analysis of pedestrian wind comfort was carried out using computational fluid dynamics (CFD) on a 3D model of the Istanbul Bilgi University Central Campus. The context of the analysis was conducted on the pedestrian activity maps within the campus between November 2021 and January 2022, including a time of strong south-westerly winds (Lodos) that hit Istanbul at the end of November/early December 2021. The activities of students and their movement patterns were observed through videos and by counting the pedestrians manually, creating a spatial map that represents pedestrian counts and their activities within the outdoor spaces. A 3D model of the campus was constructed, and CFD pedestrian wind comfort analysis was performed and compared to the pedestrian activity maps. Spatial relation analysis was performed to examine whether the pedestrian activities can be related to the Lawson wind comfort criteria map. The results show that there was generally some inconsistency between the empirically observed number of pedestrians and their activities and the wind comfort criteria map.*

DOI: 10.4018/978-1-6684-2462-9.ch010

## INTRODUCTION

The spatial, physical distribution of forms in an urban fabric affects the wind flow within the spaces between buildings, affecting the comfort of pedestrians and their activities within the space. By using Computational Fluid Dynamics (CFD) simulation software and applying the physical laws of aerodynamics, it is possible to generate a map that shows the air velocities within an urban complex. CFD simulation analysis can be performed on a mesh model of an urban area representing the physical environment of the space and its surroundings. The wind data used in the simulation is the statistical meteorological data from the weather station to the location of interest at the building site. Various wind velocities are then judged by the wind comfort criterion at a height of 1.6 meters, creating a map of different activities performed within these spaces.

Many university campuses offer a variety of daily activities within their open spaces, observation of which can facilitate beneficial modifications and improved future spatial arrangements. The strong south-westerly wind (Lodos) that hit Istanbul at the end of November 2021 was a key factor in the death of four citizens and the injury of a further 38 people, as well as numerous damages to physical structures (AFAD, 2021). This in particular highlights the necessity of considering the wind when designing open urban spaces and buildings in general, including university campuses. This research was carried out over three months from November 2021 to January 2022 with the aim of (1) investigating the relation between pedestrian activities and the Lawson wind comfort criteria map and (2) testing the suitability of open spaces in one of Istanbul's universities. Accordingly, Istanbul Bilgi University Central Campus was chosen to be the case study. Since the 2007 completion of its conversion from an electricity plant to a university campus, Bilgi Central Campus has been a host to numerous national and international exhibitions, festivals, and outdoor activities (Istanbul Bilgi University, n.d.a), while being the main campus to about 20,000 students and 1,000 academic staff and the considerable number of administrative and service staff inherent to any such large institution, as well as being utilized as a park by local residents. This study aimed to examine the relationship between activities observed during the given period and the CFD simulation map of pedestrian wind comfort analysis.

Pedestrian Wind Comfort Analysis is performed through the Computational Fluid Dynamics (CFD) application depending on 3D models to assist wind comfort and safety in urban areas. It is a form of urban space-based assessment and is most beneficial when the surrounding build environment is also accommodated in the study. Running this analysis for a particular location through CFD without considering the surrounding (or in isolation from the surrounding) built environment is sure to be less effective, as it overlooks factors which may affect the flow, direction and velocity of the wind. Wind assessment in an open area relies mainly on data provided from wind stations. Various pedestrian comfort criteria have been developed for assessing different wind frequencies and their relation to spaces and pedestrians. Lawson criteria of wind assessment depends on calculating the average hourly wind speed and categorizing these velocities into spaces where different activities are most likely to be done.

Gate count is another technique usually used by decision-makers in managing pedestrian walkability throughout the physical environment. The technique is used for the observation of pedestrian or vehicle flow, which can be used then for spatial statistical analysis. The choice of gates should cover the study area's high-, mid-, and low-flow spaces. Observers are trained to create imaginary lines after standing precisely in the location of the gate, marking a slash for every pedestrian that passes the line. Since the physical environment is one of the fundamental elements that shape pedestrians flow, the study aims to investigate the relation between pedestrian wind comfort analysis and gate counting, in order to evaluate the physical environment and its response to the walkability of Bilgi Central Campus. Accordingly, students' activities were monitored in the open spaces of the university campus with videos recorded specifically for this research project, a 3D model of the campus was created, and CFD simulation was performed.

## LITERATURE REVIEW

Numerous studies have been produced in the past years to address the effects of physical urban and building design forms and their impact on the statistics of wind (Zhang et al, 2020) (Du et al, 2017) (Elton & Shkodrani, 2020) (Serteser & Karadag, 2018). Wind flow and its velocity have been proven to change depending on either the urban form or building design form, whereby any changes of such forms lead to different wind flow and velocity results. Janssen et al. (2013a) compared four pedestrian wind comfort criteria, addressing various activities that can be done depending on the wind velocities in urban areas. This framework proves that wind velocity and flow that are affected by building forms can affect various activities performed within the urban spaces.

Performing a CFD-based pedestrian wind comfort analysis for a complex depends on three critical data inputs. These inputs are listed below: (Janssen et al., 2013a).

- A physical, virtual model of the existing or the to be built complex or building,
- The aerodynamics data, which represents both the weather data that is usually obtained from the nearest weather station to the site and the simulation that

uses the mathematical calculation of Computational Fluid Dynamics (CFD) representing the wind statistics and

- The comfort criterion

Blocken et al. (2012) overviewed the so-called 'best-practice guidelines.' Their contribution set a framework and guideline aspects that should be considered while performing CFD pedestrian wind comfort analysis. The added value of their work was to maximize the validity of the simulation by comparing the analysis results with wind speed measurements.

Zhang et al. (2020), Du et al. (2017), Elton & Shkodrani (2020) and Serteser and Karadag (2018) prove through their studies that the physical form of buildings and their height affects pedestrian comfort in terms of wind intensity. Serteser and Karadag (2018) address the necessity of considering wind comfort analysis during the architectural design of the buildings.

Through their study, Fadl and Karadelis (2013) demonstrated the importance of making wind simulations for the central campus in Coventry University after the completion of new construction projects, as these structures can deform the wind flow, causing unpleasant wind for pedestrians. These simulations are very important for the planning stage of design, because they can highlight potential problems ahead of construction. Thus, steps to mitigate against wind disruption can be mitigated against proactively. This is more financially efficient than retroactive mitigation of wind problems after construction. Using a northeast wind simulation, Bakovic et al. (2017) detected undesirable wind flows that are recognized as uncomfortable for the students within Ozyegin University campus, and recommended wind barriers to reduce the effect of wind at these areas.

Fernando et al. (2018) performed CFD wind analysis using Solid Works on a 136m tall high-rise building. The simulation was obtained at three building levels in the windiest directions according to the location's wind rose. Highrise buildings may cause some problematic and undesirable wind at the pedestrian levels, hence the importance of adding structural elements such as canopies to give better results in terms of pedestrian wind comfort (Janssen et al., 2013b). Tsichritzis and Nikolopoulou (2019) performed CFD wind comfort analysis on 24 case areas of various morphological characteristics in London, whereby the façade area ratio was found to have a linear correlation with the Building Research Establishment (BRE), reaching a comfort criteria-based activity.

## METHODOLOGY

### Case Study

Bilgi Central Campus, located in Emniyettepe neighbourhood, Eyüpsultan, Istanbul, at the edge of the Golden Horn, was chosen as a case study (see Figure 1). The site location is at latitude 41.07 and longitude 28.95. A two-way street of around 24m width separates the campus and the neighbourhood buildings from the east side. Alibeyköy creek, connected to the Golden Horn, surrounds the north and west side of the campus. The campus faces the Golden Horn from its southern side. On the side of the creek that faces the campus is the Sakarya neighbourhood. The campus buildings are a redeveloped version of a former power plant that ended its power production in 1983 (Istanbul Bilgi University, n.d.b). The campus is of an area of 118,000.0 square meters, where the total size of the buildings' footprint represents less than 22.5% of the entire campus area. More than 90,000 square meters of open green space functions as an open space for students, where students and visitors perform numerous activities. The campus is also accessible as a park for neighbourhood inhabitants during the weekends. Located in a sensitive area on the edge of Istanbul's Golden Horn, and serving as a campus for more than 20,000 students and as a park for the neighbourhood inhabitants, the site is of great importance — and the examination of its sustainability is paramount. The sustainability of the campus mostly depends on its building complex and its effect on wind flow and velocities which in turn affect the activities of pedestrians.

The wind flow can also be affected by the building patterns of the surroundings in Emniyettepe neighbourhood, which should be considered for value estimation of the comfort simulation. An AutoCAD map of the campus' surroundings containing the buildings' footprint and streets was taken from ITU Environment and Urbanism Application Research Centre. This map was modelled and used for the simulation. A diameter circle of approximately 0.82 km from the centre of the campus — including the campus buildings and their surroundings — was modelled by the present authors and used for the simulation.

### 3D Model

In order to visualize a 3D model of the existing physical situation of the campus, there was a need for an accurate map of the area. A PDF exported version of an AutoCAD file was sourced from the website of Istanbul Bilgi University (n.d.c) and then imported into AutoCAD (see Figure 2). The AutoCAD file was then imported into Revit Architecture in order to extract a 3D model of the campus. Due to the limitation of the data, the various buildings' heights were extracted — floor

*Figure 1. Bilgi University Central Campus location in Istanbul*

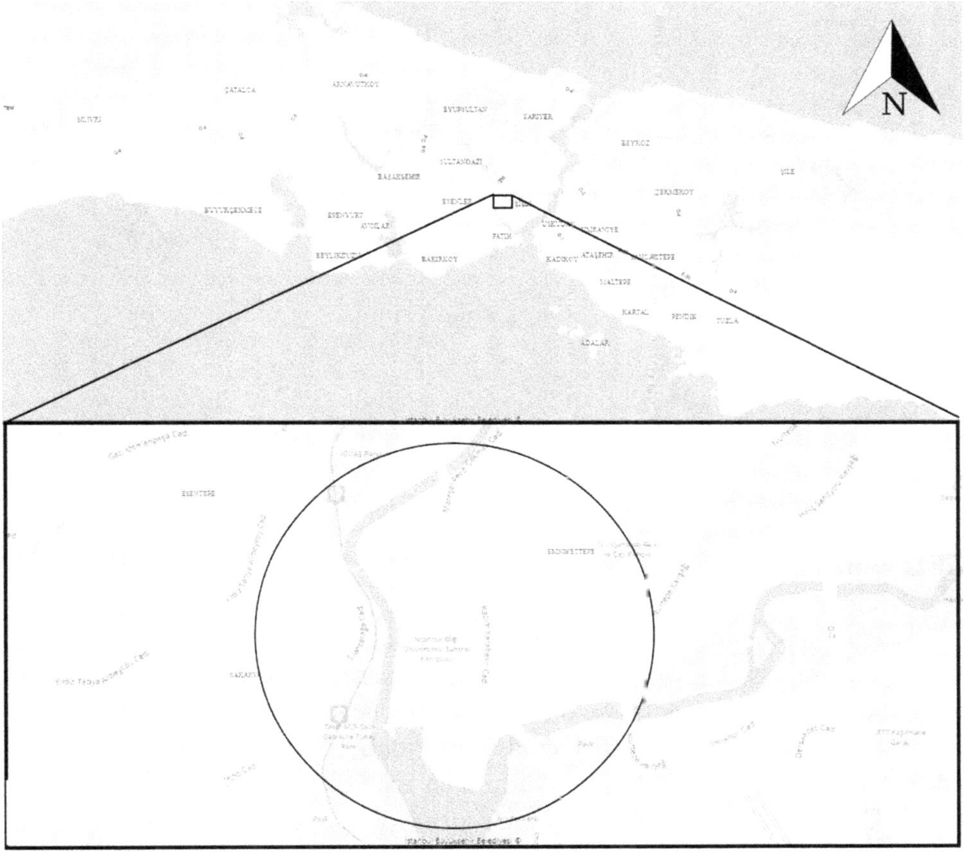


numbers were counted and an approximate height of each floor gave an approximate height of the buildings — according to observations from when the site survey, particularly from photographs taken by the researchers. An AutoCAD map of the campus' surroundings containing the buildings' footprints and streets was used as a reference for the allocation of other physical features within a circle of diameter approximately 0.82 km from the centre of the campus. This diameter — taking in the campus area and its surrounding built environment — was chosen as it ensures more accurate results in investigating the wind flow effects. The campus buildings were modelled using the "massing and site model in place" command, and the surroundings were modelled using the "in place model" and the "extraction" tools in Revit 2022 software (see Figure 3). Simple building geometries were then exported to FBX file format, but the floors and landscape were removed before exporting.

*Figure 2. Bilgi University Central Campus map with buildings' footprints*

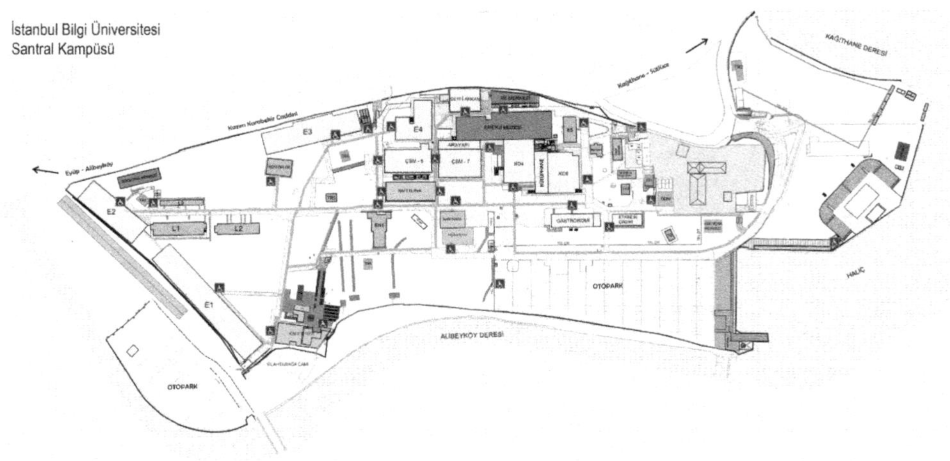


## CFD Simulation

Pedestrian wind comfort analysis can be performed using various software, depending on four data inputs. The data inputs for the simulation are (i) a 3D physical model of the existing situation or future development model, (ii) wind statistics from the nearest wind station to the site, (iii) aerodynamics data, and (iv) the comfort criterion. The software that can be used to perform such analysis can either be open source, such as OpenFOAM (OpenFOAM, n.d.), which is an opensource C++ toolbox CFD software, or alternatively commercial software or commercial cloud-based CFD apps like Simulationhub (Simulationhub, n.d.-a) or Simscale (Simscale, n.d.). This study used Simulationhub, which uses a developed version of OpenFOAM technology (Simulation, n.d.b). The reason for using commercial cloud-based software is that it provides a more user-friendly experience than open-source software alternatives. Also, it does not make any load on the user's machine; the simulations in such cases are run on servers assigned by default through the app. Jadhav et al. (2018) have represented a complete workflow of the pedestrian wind comfort analysis using the cloud-based CFD application of Simulationhub. A similar workflow was used in performing the CFD wind comfort analysis in this study, concerning modifications to the settings (user inputs).

The Revit 3D model was exported to FBX format, but the floor of the site was removed before being exported as it causes a meshing error that results in the failure of the simulation. The first factor to be taken into account when providing the app with the first data input — in this case, the 3D model — is the file format and the

*Figure 3. Bilgi University Central Campus 3D mass model using Revit 2022*

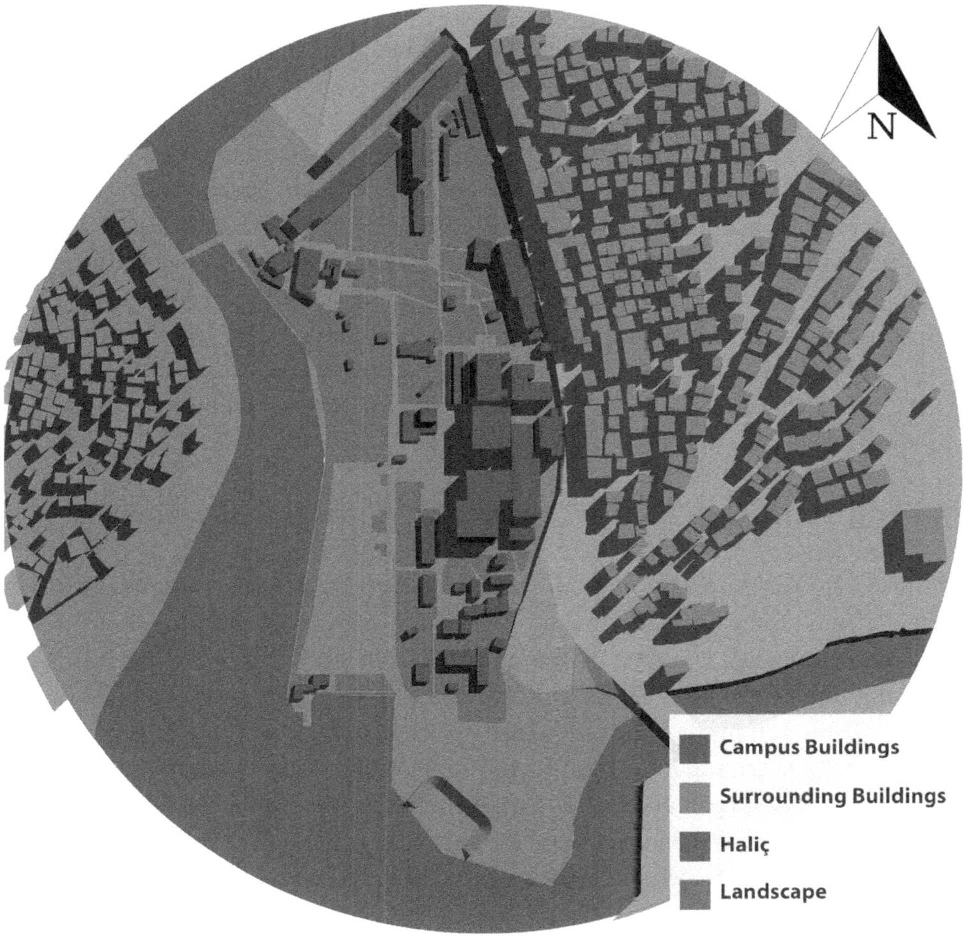


geometries' specifications. The 3D model has to be aligned with the coordinates of the app, where the orientation of the north has to be considered before modelling or rotated to face north — after modelling but before exporting.

The location of the site is at latitude 41.07, longitude 28.95. The nearest wind station is Istanbul Ataturk meteorological station (TUR_Istanbul.170600_IWEC), from which weather data of this simulation was taken. Figure 4 and Table 1 show the wind distribution based on historical wind data at the station. The distance from the station to the site is approximately 15km. For an accurate result, wind condition from 32 compass direction was collected, and a total of 32 simulations of various direction were performed and used to judge the wind comfort.

Wind exposure should also be taken into consideration when running CFD simulations. Since the wind velocities are measures taken from the nearest metrological stations and measured in an environment that is different from the case study environment, defining the type of environment surrounding the case study is crucial to attaining accurate wind velocity measurements using the physical laws of aerodynamics. Considering that 0 is north and 90 is east, Figure 5 and Table 2 show the defined exposures. The final input of the simulation is the comfort criteria.

Numerous criteria have been developed to estimate the relation between wind velocity and pedestrians' comfort and safety in urban areas. One such criteria is Lawson criteria (1977). The criteria define classifications of activities and safety depending on wind velocities and exceedance percentages. The simulation of this study was run according to the given inputs. Figure 6 shows the wind simulation results using the cloud-based CFD application of Simulationhub according to Lawson criteria.

## Walking Tour, Pedestrian Counts and Activity Observations

The research was carried out by a team of researchers and several volunteers — students from a range of departments — who assisted in the observation work on the campus. In the first week of December 2021, the observers were asked to take a walking tour throughout Bilgi Central Campus. The main aim of the tour was to (1) observe the campus in company with the current university students to know the building and get their input regarding activities that take place within the campus, (2) take pictures of the physical structures and estimate an approximate height according to the floor numbers and their approximate height, (3) observe the study area's high-, mid-, and low-flow spaces, and (4) choose the gate positions on a printed map of the site and allocate appropriate locations for the observers where the view range of their cameras could capture the pedestrians within their assigned axes.

The gate count technique is a quantitative method used to examine the frequency of pedestrians or vehicles, where certain gates are distributed within the principal axes of traffic in a particular site. The gate choice should cover the study area's axes of various traffic densities. A total number of 10 gates were chosen along different axes within the university campus and a certain walking tour was assigned by the researchers to be followed by the observers, see Figure 7. The observers were asked to record videos from which pedestrians could be counted. Firstly, the campus video footage was recorded on the 2nd of December 2021 (during the south-westerly wind), whereby an observer was asked to follow the given walking tour and document the environment from 12:30-13:00. The aim of the walk was to concentrate on recording the students' activities and gatherings during a period of strong wind within the campus' open space. Observers were also asked to concentrate on the open spaces

*Figure 4. Wind distribution (Wind Rose) based on historical wind data at the weather station*

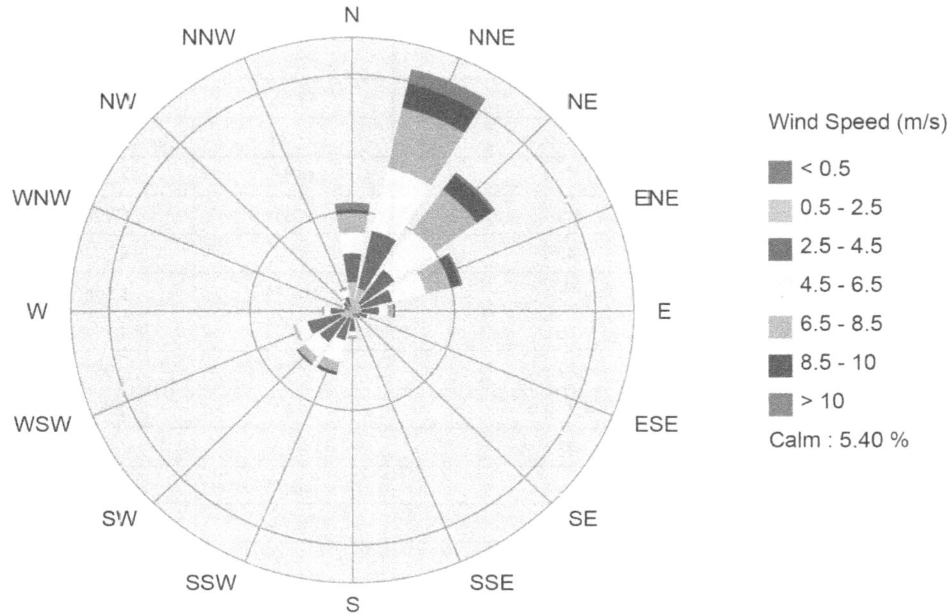


within the campus that they deemed least exposed to the strong wind, then to mark these observations on the map.

Secondly, other videos were taken over two days, including 1 weekday (29 December 2021) and one weekend (15th January 2022). An observer was asked to follow the same walking tour and document the environment during three intervals within each counting day, 09:30-10:00, 16:30-17:00 and 18:30-19:00. Manual counting of pedestrians from both directions was then carried out from the video footage, whereby an imaginary line was created and a slash was then marked for every pedestrian that passed the line. Pedestrian activities were also taken into account, with observers asked to focus the camera angle on pedestrians. Such videos can create a map showing pedestrian activities and their numbers. A total time of 80 minutes of video was used to create a database and maps for the pedestrian counting and activities (see Figure 9, Figure 10 and Figure 11).

## Workflow and Spatial Relationship

The general workflow of the study can be seen in Figure 8. After performing the CFD simulation according to Lawson criteria and the above data inputs, a map generated

*Table 1. Wind conditions of 32 directions based on historical wind data at the weather station*

| Condition No. | Simulation Velocity (m/s) | Direction (°) | Frequency |
|---|---|---|---|
| 0 | 5.903 | 45 | 15.97% |
| 1 | 5.5 | 11.25 | 11.53% |
| 2 | 6.08 | 33.75 | 11.039% |
| 3 | 5.789 | 56.25 | 7.215% |
| 4 | 5.9 | 22.5 | 6.998% |
| 5 | 4.405 | 225 | 6.244% |
| 6 | 4.919 | 213.75 | 3.596% |
| 7 | 3.836 | 236.25 | 3.333% |
| 8 | 4.836 | 67.5 | 3.037% |
| 9 | 4.719 | 202.5 | 2.58% |
| 10 | 3.427 | 247.5 | 2.146% |
| 11 | 4.197 | 348.75 | 2.123% |
| 12 | 4.756 | 78.75 | 1.678% |
| 13 | 1.506 | 0 | 1.507% |
| 14 | 3.98 | 191.25 | 1.404% |
| 15 | 3.501 | 315 | 1.37% |
| 16 | 3.86 | 90 | 1.267% |
| 17 | 3.999 | 326.25 | 1.142% |
| 18 | 4.612 | 337.5 | 1.142% |
| 19 | 3.895 | 270 | 1.13% |
| 20 | 3.65 | 258.75 | 1.073% |
| 21 | 3.176 | 135 | 1.05% |
| 22 | 3.718 | 101.25 | 0.947% |
| 23 | 3.296 | 123.75 | 0.879% |
| 24 | 3.028 | 112.5 | 0.811% |
| 25 | 3.447 | 180 | 0.799% |
| 26 | 3.629 | 292.5 | 0.548% |
| 27 | 3.282 | 281.25 | 0.502% |
| 28 | 4.272 | 303.75 | 0.457% |
| 29 | 2.82 | 146.25 | 0.4% |
| 30 | 2.779 | 168.75 | 0.377% |
| 31 | 2.804 | 157.5 | 0.308% |

from 32 steady-state maps representing five activities and safety classifications was created. In addition, 80 minutes of video footage taken from the defined ten gates over both weekends and weekdays and another 10-minute video taken at a

*Figure 5. Wind exposure*

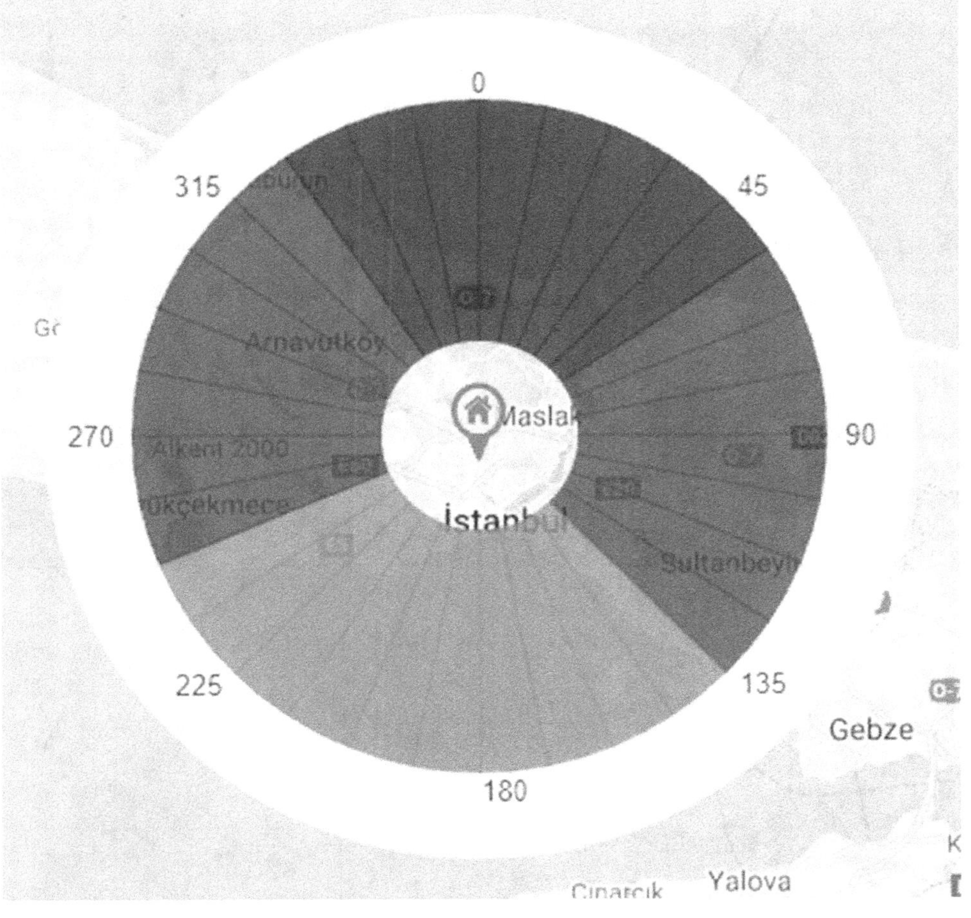


*Table 2. Wind exposure*

| From | To | Exposure |
|---|---|---|
| 326.25 | 56.25 | Exposed to Suburban |
| 56.25 | 135 | Exposed to City |
| 135 | 247.5 | Exposed to Sea |
| 247.5 | 326.25 | Exposed to City |

certain day during the south-westerly wind were defined as inputs for the activities database. Activities databases were then spatially represented on the university map

*Figure 6. Wind simulation results using the cloud-based CFD application of Simulationhub. Map on the right, 3D shot on the upper left and southwestern wind flow 3D shot on the lower left*

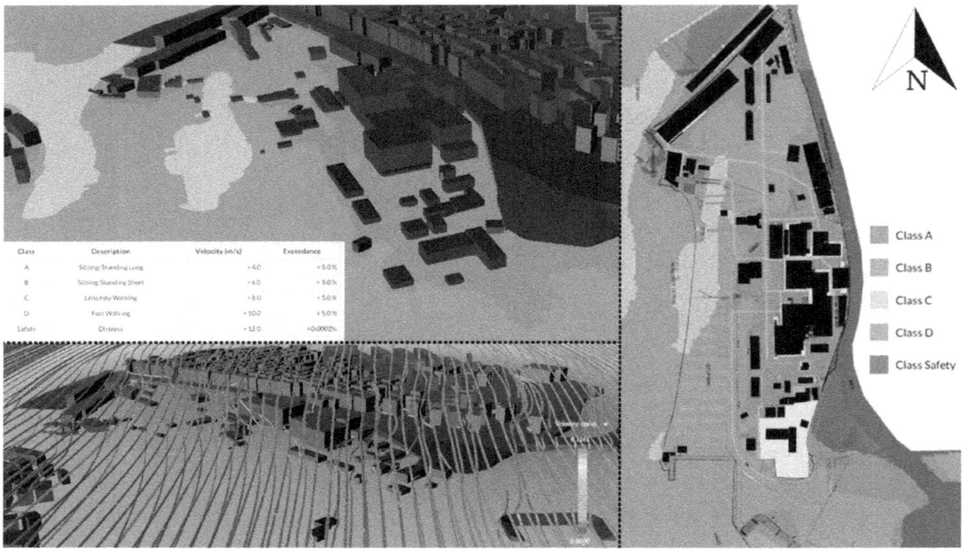


*Figure 7. Bilgi University central campus walking tour and pedestrian counts' gate*

*Figure 8. The general workflow of the study*

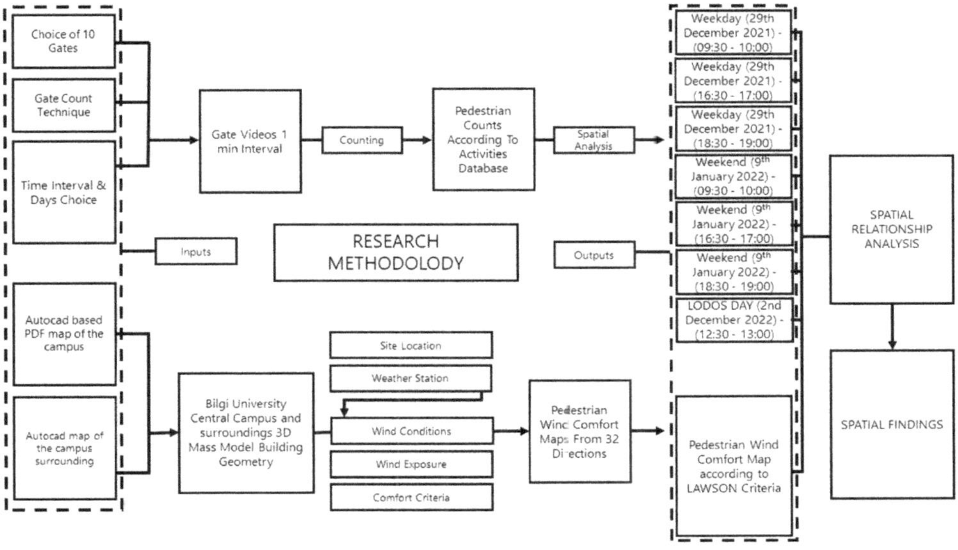


and a total of 7 maps representing pedestrian activities with respect to counts were created (see Figure 9, Figure 10 and Figure 11). In order to find a spatial relationship between the pedestrian wind comfort analysis and the pedestrian activity maps, spatial correlation analysis was performed.

The spatial relationship is used to address the relations between maps that have spatial data attributes. For this study, the simulation map produced from CFD pedestrian wind comfort analysis was categorized, and the pedestrian counts were also based on the same categorization of activities. Thus, it was clear that the suitability assessment would depend on allocating the gates and observing the correspondence of the gate on the activity map. As stated above, according to the simulation map, the lower wind velocity areas facilitate a broader range of activities than those of higher velocity. It follows those areas that are suited to sitting or standing for a long period of time are thus suitable for the same kinds of activity over a shorter period – but not the reverse. The challenge is to recognize exceptions in practice to the CFD categorizations of behaviour and environmental conditions and to try to account for them on the observation map. This is the technique that has been used in the assessment in this study.

*Figure 9. Activity map on Weekday (29 December 2021)*

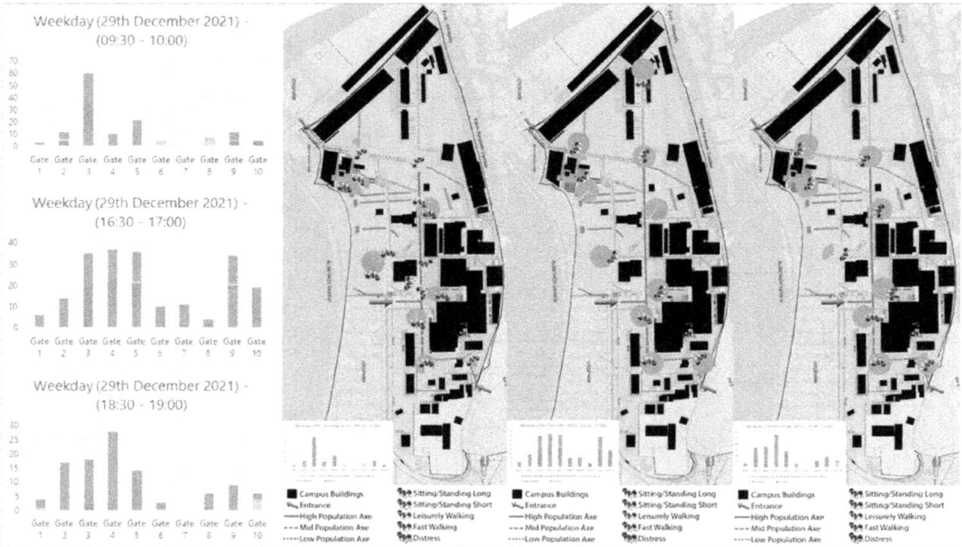


## FINDINGS AND DISCUSSION

In order to find the spatial relationship between the pedestrian activities and the CFD pedestrian wind comfort analysis, the activity map was put on top of the CFD map, (see Figure 15).

At Gates 1 and 2, the space was found to be suitable for a range of activities, since it is ideal for sitting/standing for a short time. At the same time, no activities were observed at these two gates, possibly because there is no urban furniture provided. Other activities are infrequent at these gates, although according to the simulation map, they fit the criteria for an actively used space. Gate 3 and Gate 4 were deemed to be ideal for standing/sitting for a longer duration. These gates are the most crowded, due to their being at the main entrance to the core of the campus. Most kinds of activities were seen to be performed within proximity of these gates. Even on the day during the south-westerly wind, students chose to utilize these gates, but in areas that seemed relatively less exposed to wind flow (see Figure 12). Notably, much of the campus was relatively empty on this day. The northern side of the gates was more crowded than the southern side, where urban furniture had been placed but mostly unused. According to Figure 15 below, groups of 18 and 13 pedestrians respectively were observed sitting for a long time at Gates 3 and 4 on both 2 December 2021 and 29 December 2021. According to the CFD analysis, both are located within an area that should be suitable for standing/sitting for a short

*Figure 10. Activity map at Weekend (15 January 2022)*

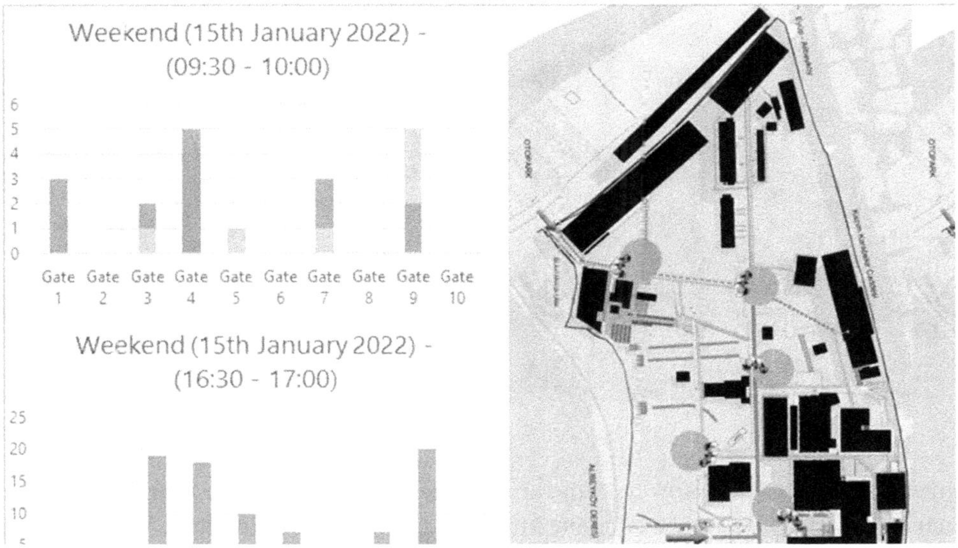


*Figure 11. Activity map at the day during the south-westerly wind (2 December 2022 during 12:30 – 13:00)*

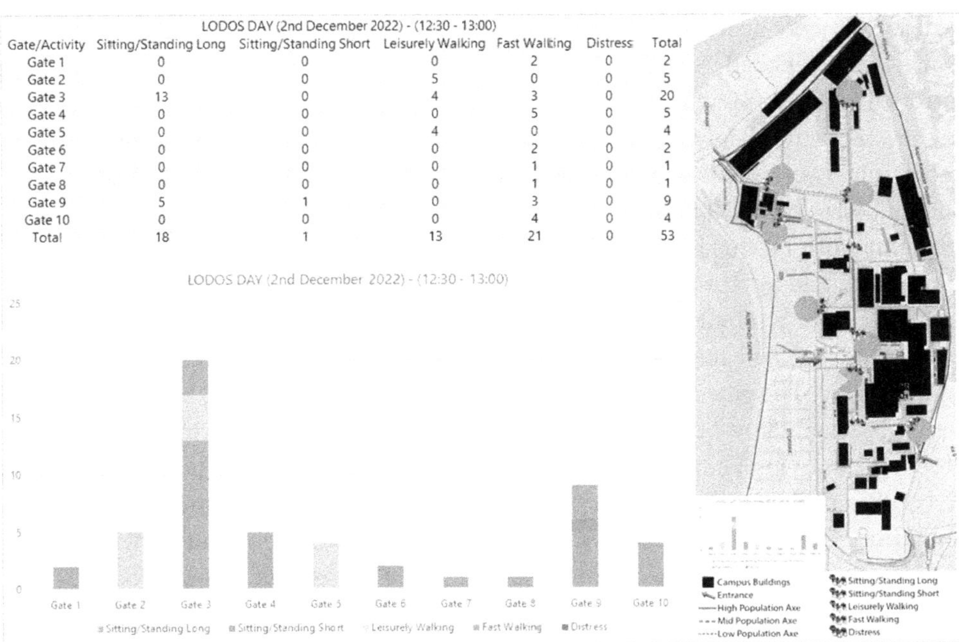

LODOS DAY (2nd December 2022) - (12:30 - 13:00)

| Gate/Activity | Sitting/Standing Long | Sitting/Standing Short | Leisurely Walking | Fast Walking | Distress | Total |
|---|---|---|---|---|---|---|
| Gate 1 | 0 | 0 | 0 | 2 | 0 | 2 |
| Gate 2 | 0 | 0 | 5 | 0 | 0 | 5 |
| Gate 3 | 13 | 0 | 4 | 3 | 0 | 20 |
| Gate 4 | 0 | 0 | 0 | 5 | 0 | 5 |
| Gate 5 | 0 | 0 | 4 | 0 | 0 | 4 |
| Gate 6 | 0 | 0 | 0 | 2 | 0 | 2 |
| Gate 7 | 0 | 0 | 0 | 1 | 0 | 1 |
| Gate 8 | 0 | 0 | 0 | 1 | 0 | 1 |
| Gate 9 | 5 | 1 | 0 | 3 | 0 | 9 |
| Gate 10 | 0 | 0 | 0 | 4 | 0 | 4 |
| Total | 18 | 1 | 13 | 21 | 0 | 53 |

*Figure 12. Gate 4 at the day during the south-westerly wind*

time. Contrary to this, however, the areas were found to be used by pedestrians for longer periods. This can be attributed to the cafeteria and shaded light structure that exist near the entrance and at the core of the open space of the university campus.

Gate 5 was seen to be one of the most crowded gates, where most activities were observed — including sitting/standing for a long and shorter period and leisurely or fast walking. The gate is located beside buildings that are frequently used by students. On the comfort map, the space was found to be suitable for sitting/standing for only a short time — less so for longer periods. On weekdays, groups of 9 or 10 pedestrians were observed sitting for a long time at each of the observation intervals. During the weekend, 3 pedestrians were observed sitting for a long time at Gate 5 from 16:30-17:00, whereas no pedestrians were observed sitting/standing there at the other intervals. On the day during the south-westerly wind, only walking pedestrians were seen — no sitting/standing activities were observed. Gate 5 was understood to be a potentially favourable space for students to gather. However, it is currently unfit for this purpose. The addition of certain amenities, such as a covered and semi-enclosed space that would be able to protect its users from strong winds and heavy rain without obscuring the view, would benefit the area.

Gate 6 is located at a distribution path that branches out to a number of buildings and the upper left gate. The gate has a relatively high traffic flow, but no activities were observed there other than walking. The gate is surrounded by underutilized green spaces of high potential, which could be used for several functions by students. According to the CFD, it is a space that could be suitable for sitting/standing for a long time. However, the space lacks any urban furniture that might invite students to view it differently. From our observations, Gate 7 was considered to be the least crowded route. This was supported by the views of the observers, who stated that

*Figure 13. Gate 3*

it is generally the least used route within the campus. Gate 8 is located on the route that leads to one entrance to the campus, and was classified in the CFD as suitable for sitting/standing for a short time. No sitting activities were observed within the gate, and it also lacks any urban furniture.

Although Gate 9 was not within the CFD category of a suitable space for standing/sitting for a long period, it was seen in the observations to be one of the most crowded gates. All manner of activities within the framework were observed in its space. The gate is located next to a cafeteria and contains a covered and semi-enclosed area that can protect users from strong winds and heavy rain (see Figure 14). On both weekdays and weekends, smaller and larger groups of pedestrians were seen to gather there for shorter and longer periods. After Gates 3 and 4, Gate 9 was one of the most utilized gates on the day during the south-westerly wind, with a number of students observed sitting and standing there. Although the gate falls inside the CFD boundaries of spaces that might not be suitable for sitting/standing for a long time, the existence of the covered area made the space usable in various environmental conditions without closing the surrounding views.

Gate 10 is located on the eastern side, and the gate falls into the leisurely walking boundaries according to Lawson criteria. Smaller groups of pedestrians were found standing/sitting for longer and shorter times at this gate, whereas on the day during the south-westerly wind, the gate was found to be unused. The gate could potentially be used for sitting/standing for a short time. This would require the additions of a covered and semi-enclosed area that is able to protect its users from strong winds and heavy rain without obscuring the view.

*Figure 14. Semi-enclosed space that is able to protect users from strong winds and heavy rain*

Overall, there was generally some inconsistency between the empirically observed number of pedestrians and their activities and the CFD map. In some of the gates that were classified as suitable for a range of activities — since they were deemed ideal for sitting/standing for a long time — little to no activities were observed (for example, Gates 1, 2 and 7). In contrast, gates that were expected to be mainly suited to just leisurely walking were seen in the survey to be the most crowded gates, where a broader range of activities were observed (for example, Gate 9). It was also noted that some gates that have the potential — in terms of their suitability to wind comfort criteria — to be used at the weekend or weekdays, were found to be unused, so that no activities were observed there other than walking.

On the observation day of the south-westerly wind, a diversity of activities was observed at Gates 3 and 9, while at the rest of the gates the activities were limited to walking. It was noted that pedestrians chose semi-closed places to avoid the wind at the gates where such an option was available, whereas the rest of the gates were used only transiently for access from one point to another. Unsurprisingly, less pedestrian activity was observed in areas with high south-westerly wind, while busier and more diversified pedestrian activity was detected in areas where the south-westerly wind caused less disturbance. We can reasonably conclude that this demonstrates how wind density and velocity affect pedestrian comfort conditions and their activities.

*Figure 15. Activity map placed on top of CFD map*

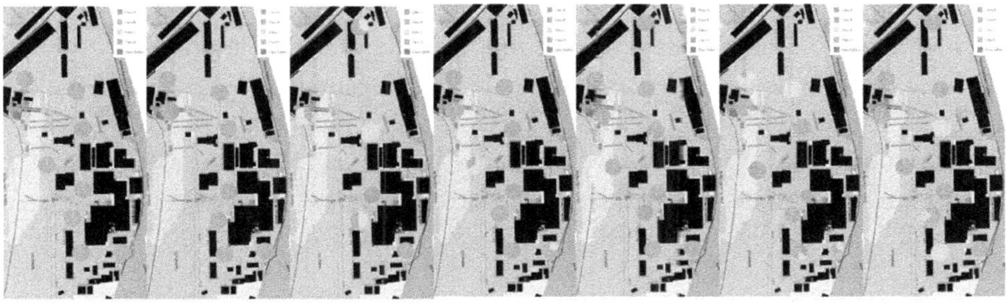

| Weekday (29th December 2021) - (08:30 - 10:00 | | | | | kday (29th December 2021) - (16:30 - 1 | | | | | kday (29th December 2021) - (18:30 - 1 | | | | | ekend (15th January 2022) - (09:30 - 10 | | | | | ekend (15th January 2022) - (16:30 - 17 | | | | | ekend (15th January 2022) - (18:30 - 19:15 | | | | | Day (2nd December 2022) - (12:30 - 1 | | | | | | |
|---|---|---|---|---|---|---|---|---|---|---|---|---|---|---|---|---|---|---|---|---|---|---|---|---|---|---|---|---|---|---|---|---|---|---|---|---|
| te / Acti | Standi | Standi | rely Wa | t Walk | Distress | Standi | Standi | rely Wa | t Walk | Distress | Standi | Standi | rely Wa | t Walk | Distress | Standi | Standi | rely Wa | t Walk | Distress | Standi | Standi | rely Wa | t Walk | Distress | Standi | Standi | rely Wa | t Walk | Distress | Standi | Standi | rely Wa | t Walk | Distress | Class |
| Gate 1 | 0 | 0 | 1 | 2 | 0 | 0 | 0 | 0 | 6 | 0 | 0 | 0 | 1 | 3 | 0 | 0 | 0 | 0 | 3 | 0 | 0 | 0 | 0 | 3 | 0 | 0 | 0 | 0 | 7 | 0 | 0 | 0 | 0 | 2 | 0 | Sitting/Standing Long |
| Gate 2 | 0 | 4 | 2 | 5 | 0 | 0 | 0 | 0 | 14 | 0 | 0 | 0 | 0 | 17 | 0 | 0 | 0 | 0 | 0 | 0 | 0 | 0 | 0 | 1 | 0 | 0 | 0 | 0 | 2 | 0 | 0 | 0 | 5 | 0 | 0 | Sitting/Standing Long |
| Gate 3 | 25 | 14 | 0 | 26 | 0 | 0 | 19 | 0 | 16 | 0 | 0 | 2 | 0 | 16 | 0 | 0 | 0 | 1 | 1 | 0 | 0 | 3 | 0 | 16 | 0 | 0 | 0 | 0 | 20 | 0 | 13 | 0 | 4 | 3 | 0 | Sitting/Standing Short |
| Gate 4 | 9 | 0 | 0 | 1 | 0 | 10 | 12 | 0 | 15 | 0 | 18 | 0 | 0 | 10 | 0 | 0 | 4 | 0 | 1 | 0 | 0 | 5 | 0 | 13 | 0 | 4 | 0 | 0 | 1 | 0 | 0 | 0 | 0 | 5 | 0 | Sitting/Standing Long |
| Gate 5 | 9 | 0 | 0 | 12 | 0 | 10 | 11 | 1 | 14 | 0 | 8 | 3 | 0 | 5 | 0 | 0 | 0 | 1 | 0 | 0 | 3 | 0 | 0 | 7 | 0 | 0 | 0 | 0 | 3 | 0 | 0 | 0 | 4 | 0 | 0 | Sitting/Standing Short |
| Gate 6 | 0 | 0 | 4 | 0 | 0 | 0 | 0 | 0 | 10 | 0 | 0 | 0 | 0 | 3 | 0 | 0 | 0 | 0 | 0 | 0 | 0 | 0 | 0 | 7 | 0 | 0 | 0 | 0 | 2 | 0 | 0 | 0 | 0 | 2 | 0 | Sitting/Standing Short |
| Gate 7 | 0 | 0 | 0 | 0 | 0 | 0 | 0 | 0 | 11 | 0 | 0 | 0 | 0 | 0 | 0 | 0 | 0 | 1 | 2 | 0 | 0 | 0 | 0 | 0 | 0 | 0 | 0 | 0 | 0 | 0 | 0 | 0 | 0 | 1 | 0 | Sitting/Standing Long |
| Gate 8 | 0 | 0 | 6 | 1 | 0 | 0 | 1 | 0 | 3 | 0 | 0 | 0 | 0 | 6 | 0 | 0 | 0 | 0 | 0 | 0 | 0 | 0 | 0 | 7 | 0 | 0 | 3 | 0 | 3 | 0 | 0 | 0 | 0 | 1 | 0 | Sitting/Standing Short |
| Gate 9 | 2 | 2 | 1 | 7 | 0 | 18 | 2 | 1 | 13 | 0 | 4 | 0 | 0 | 5 | 0 | 2 | 0 | 8 | 0 | 0 | 0 | 7 | 0 | 13 | 0 | 0 | 0 | 0 | 0 | 0 | 5 | 1 | 0 | 5 | 0 | Leisurely Walking |
| Gate 10 | 1 | 8 | 0 | 0 | 0 | 4 | 0 | 0 | 15 | 0 | 0 | 0 | 4 | 2 | 0 | 0 | 0 | 0 | 0 | 0 | 0 | 0 | 0 | 0 | 0 | 0 | 0 | 0 | 3 | 0 | 0 | 0 | 0 | 4 | 0 | Leisurely Walking |

## CONCLUSION

CFD pedestrian wind comfort analysis and other urban analysis are a crucial aspect of architectural and urban design and should be integrated at the earliest feasible stages of the design process. To achieve maximum potential in terms of the diverse usage of space, it is recommended to take account of wind simulation analysis. The CFD wind simulation is found to be user friendly, but such applications are commercial and highly priced. 3D models offer potential advantages in the visualization of projects to clients and for running various analyses to test a project's suitability before execution — which directly influences the project budget. CFD pedestrian wind comfort analysis is recommended for use on existing structures in order to analyse or find relations between people and spaces. Hence it can be of great importance for suitable and reliable future modifications. Conducting this sort of analysis can help identify the degree of risk, and enable to designers to develop strategies to mitigate against the risk during designing phases or even for existing structures.

CFD pedestrian wind comfort analysis is also of great potential as a tool that can be used in the allocation and design of urban furniture in open spaces and public areas. Lawson criteria of wind comfort analysis offers a relation between open spaces and activities that are categorized according to the wind velocities and in this study was useful in assessing the cause-effect relation of pedestrians and open spaces in terms of wind velocity. The CFD pedestrian wind comfort analysis performed in this study using commercial software lacks data regarding detailed wind velocity maps that can be used for regression, which can be deducted using Opensource software.

On the other hand, Opensource software lacks a user-friendly experience, potentially making the analysis more difficult for architects and urban designers. There is a need to develop Opensource software and provide it with a more user-friendly interface that can be used on a broader scale in the designing stages. Additionally, the lack of video tutorials that show the step-by-step procedure of the analysis in non-commercial software may have a negative impact on the duration of any study, leading to the shortcomings in the analysis at the designing and later stages.

## REFERENCES

AFAD. (2021). *About*. Retrieved December 6, 2021, from https://www.afad.gov.tr/ulke-genelindeki-olumsuz-hava-kosullari-hk-basin-bulteni---3-30112021-saat-0825-itibariyla

Bakovic, M., Siddiqui, F., Başol, M. A., & Göçer, Ö. (2017). Outdoor Thermal Comfort Analysis at a Sustainable University Campus. In *International Symposium to Promote Innovation & Research in Energy Efficiency-INSPIRE Symposium* (*Vol. 27*). Academic Press.

Blocken, B., Janssen, W. D., & van Hooff, T. (2012). CFD simulation for pedestrian wind comfort and wind safety in urban areas: General decision framework and case study for the Eindhoven University campus. *Environmental Modelling & Software*, *30*, 15–34. doi:10.1016/j.envsoft.2011.11.009

Du, Y., Mak, C. M., Liu, J., Xia, Q., Niu, J., & Kwok, K. C. (2017). Effects of lift-up design on pedestrian level wind comfort in different building configurations under three wind directions. *Building and Environment*, *117*, 84–99. doi:10.1016/j.buildenv.2017.03.001

Elton, H. A. L. A., & Shkodrani, N. (2020). Design for wind comfort. The CFD assessment over a future outdoor public space. *Academic Platform Journal of Natural Hazards and Disaster Management*, *1*(1), 13–24.

Fadl, M. S., & Karadelis, J. (2013). CFD simulation for wind comfort and safety in urban area: A case study of Coventry University central campus. *International Journal of Architecture, Engineering and Construction*, *2*(2), 131-143.

Fernando, S., Fernando, S., & Mendis, P. (2018, December). Pedestrian wind comfort study using computational fluid dynamic (CFD) simulation. In *International Conference on Sustainable Built Environment* (pp. 323-339). Springer.

Istanbul Bilgi University. (n.d.a). *About santralistanbul*. Retrieved November 27, 2021, from https://www.santralistanbul.org/tr/hakkinda/

Istanbul Bilgi University. (n.d.b). *About Silahtarağa Power Plant*. Retrieved November 27, 2021, from https://www.santralistanbul.org/en/silahtaraga-power-plant/

Istanbul Bilgi University. (n.d.c). *About İstanbul Bilgi University Santral Campus map*. Retrieved November 12, 2021, from https://www.bilgi.edu.tr/upload/santralistanbul-campus-accessible-campus-map/

Isyumov, N., & Davenport, A. G. (1975). The ground level wind environment in built-up areas. In *Proceedings of the 4th International Conference on Wind Effects on Buildings and Structures* (pp. 403-422). Academic Press.

Jadhav, S., Mali, V., Ramachandran, P. K., & Rane, C. (2018). *Large Scale Modelling of Wind Comfort and Safety Using 'Pedestrian Comfort Analysis'-A Cloud-based App Crafted for Architects*. Centre for Computational Technologies.

Janssen, W., Blocken, B., & van Hooff, T. (2013-b). Use of CFD simulations to improve the pedestrian wind comfort around a high-rise building in a complex urban area. *13th Conference of International Building Performance Simulation Association*.

Janssen, W. D., Blocken, B., & van Hooff, T. (2013-a). Pedestrian wind comfort around buildings: Comparison of wind comfort criteria based on whole-flow field data for a complex case study. *Building and Environment*, *59*, 547–562. doi:10.1016/j.buildenv.2012.10.012

Lawson, T. V. (1977). The effects of wind on people in the vicinity of buildings. Academic Press.

OpenFOAM. (n.d.). *About OpenFOAM*. Retrieved December 21, 2021, from https://www.openfoam.com/

Serteser, N., & Karadag, I. (2018). Design for improving pedestrian wind comfort: A case study on a courtyard around a tall building. *Architectural Science Review*, *61*(6), 492–499. doi:10.1080/00038628.2018.1492899

Simscale. (n.d.). *About Simscale*. Retrieved December 21, 2021, from https://www.simscale.com/

Simulationhub. (n.d.a). *About Simulationhub*. Retrieved December 21, 2021, from https://www.simulationhub.com/

Simulationhub. (n.d.b). *About Simulationhub credits*. Retrieved December 21, 2021, from https://www.simulationhub.com/credits

Tsichritzis, L., & Nikolopoulou, M. (2019). The effect of building height and façade area ratio on pedestrian wind comfort of London. *Journal of Wind Engineering and Industrial Aerodynamics*, *191*, 63–75. doi:10.1016/j.jweia.2019.05.021

Zhang, X., Weerasuriya, A. U., Zhang, X., Tse, K. T., Lu, B., Li, C. Y., & Liu, C. H. (2020, December). Pedestrian wind comfort near a super-tall building with various configurations in an urban-like setting. *Building Simulation*, *13*(6), 1385–1408. doi:10.100712273-020-0658-6 PMID:32837689

# Compilation of References


Abbas, I. F., Purwaningrum, P., & Indrawati, D. (2021). Evaluation of waste management in tourism area of Luar Batang Village, outside area of the Historic Old Jakarta Kota, Penjaringan District, North Jakarta. *IOP Conference Series. Earth and Environmental Science*, *737*(1), 012012. doi:10.1088/1755-1315/737/1/012012

Abdelaziz, L. (2021). Heliopolis city, a cultural political interpretation. *AGORA CAUMME 2020-Mediterranean: Between expansion and regeneration.*

Abdou, A., Abd Elgawad, I., & Tarek, T. (2016, July). Green Infrastructure to achieve Sustainability in Urban Design: Nile Corridor in Great Cairo as a Case Study. *Journal of Al Azhar University of Engineering Sector.*, *11*(40), 1055–1066. doi:10.21608/auej.2016.19410

Abusaada, H. (2017). The untold story of the urban design practice of new cities in Egypt. In H. Elsharkawy, S. Zahiri, & J. Clough (Eds.), *In the proceeding of the international conference for sustainable design of the built environment* (pp. 787–798). SDBE.

Abusaada, H. (2020). Strengthening the affectivity of atmospheres in urban environments: The toolkit of multi-sensory experience. *Archnet-IJAR*, *14*(3), 379–392. doi:10.1108/ARCH-03-2020-0039

Abusaada, H., & Elshater, A. (2019). Effects of urban atmospheres on changing attitudes of crowded public places: An action plan. *International Journal of Community Well-Being*, *3*(2), 109–159. doi:10.100742413-019-00042-w

Abusaada, H., & Elshater, A. (2020). Urban design assessment tools: A model for exploring atmospheres and situations. *Proceedings of the Institution of Civil Engineers. Urban Design and Planning*, *173*(6), 238–255. doi:10.1680/jurdp.20.00025

Abusaada, H., & Elshater, A. (2021a). Building sustainable habitats for free-roaming cats in public spaces: A systematic literature review. *Journal of Applied Animal Welfare Science*, 1–14. doi:10.1080/10888705.2021.2014840 PMID:34957863

Abusaada, H., & Elshater, A. (2021a). Competitiveness, distinctiveness and singularity in urban design: A systematic review and framework for smart cities. *Sustainable Cities and Society*, *68*, 1–12. doi:10.1016/j.scs.2021.102782

Abusaada, H., & Elshater, A. (2021b). Effect of people on placemaking and affective atmospheres in city streets. *Ain Shams Engineering Journal, 12*(3), 3389–3403. doi:10.1016/j.asej.2021.04.019

Abusaada, H., & Elshater, A. (2021b). Revealing distinguishing factors between Space and Place in urban design literature. *Journal of Urban Design, 26*(3), 319–340. doi:10.1080/13574809.2020.1832887

Abusaada, H., & Elshater, A. (2021c). Improving visitor satisfaction in Egypt's Heliopolis historical district. *Journal of Engineering and Applied Sciences (Asian Research Publishing Network), 68*(19), 1–22. doi:10.118644147-021-00022-y

Abutaleb, K., Ngie, A., Darwish, A., Ahmed, M., Arafat, S., & Ahmed, F. (2015). Assessment of urban heat island using remotely sensed imagery over Greater Cairo, Egypt. *Advances in Remote Sensing, 04*(01), 35–47. doi:10.4236/ars.2015.41004

Adhikari, S. (2018, March 6). *Urban heat island, its effects and mitigation measures*. Retrieved from Public Health Notes: https://www.publichealthnotes.com/

AFAD. (2021). *About*. Retrieved December 6, 2021, from https://www.afad.gov.tr/ulke-genelindeki-olumsuz-hava-kosullari-hk-basin-bulteni---3-30112021-saat-0825-itibariyla

Ahmed, M., Scerbo, M., Izzo, P., Parrilli, M., Coccia, F., Ganga, V., & Anilir, S. (2009). A Community-based Waste Management System for the Historic Centre of Naples. *Journal of Asian Architecture and Building Engineering, 8*(2), 363–370. doi:10.3130/jaabe.8.363

Ahmed, T., Moeinaddini, M., Almoshaogeh, M., Jamal, A., Nawaz, I., & Alharbi, F. (2021). A New Pedestrian Crossing Level of Service (PCLOS) Method for Promoting Safe Pedestrian Crossing in Urban Areas. *International Journal of Environmental Research and Public Health, 18*(16), 1–18. doi:10.3390/ijerph18168813 PMID:34444563

Alattar, D. A., & Furlan, R. (2017). Urban regeneration in Qatar: A comprehensive planning strategy for the transport oriented development of Al-Waab. *Journal of Urban Regeneration and Renewal, 11*(2), 168–193.

Alchapar, N. L., & Cantón, N. A. (2016). Classification of building materials used in the urban envelopes according to their capacity for mitigation of the urban heat island in semiarid zones. *Energy and Building, 69*, 22–32. doi:10.1016/j.enbuild.2013.10.012

Algeciras, J. A. R., Tablada, A., & Matzarakis, A. (2018). Effect of asymmetrical street canyons on pedestrian thermal comfort in the warm-humid climate of Cuba. *Theoretical and Applied Climatology, 133*(3-4), 663–679. doi:10.100700704-017-2204-8

Ambrosini, D., Galli, G., Mancini, B., Nardi, I., & Sfarra, S. (2014). Evaluating mitigation effects of urban heat islands in a historical small center with the ENVI-Met® climate model. *Sustainability, 6*(10), 7013–7029. doi:10.3390u6107013

Anciaes, P. R., & Jones, P. (2018). Estimating preferences for different types of pedestrian crossing facilities. *Transportation Research Part F: Traffic Psychology and Behaviour, 52*, 222–237. doi:10.1016/j.trf.2017.11.025

Anguelovski, I., Shi, L., Chu, E., Gallagher, D., Goh, K., Lamb, Z., Reeve, K., & Teicher, H. (2016). Equity impacts of urban land use planning for climate adaptation: Critical perspectives from the Global North and South. *Journal of Planning Education and Research*, *36*(3), 333–348. doi:10.1177/0739456X16645166

Antonini, G., Bierlaire, M., & Weber, M. (2005). Discrete choice models of pedestrian walking behavior. *Transportation Research Part B: Methodological*, *40*(8), 667–687. doi:10.1016/j.trb.2005.09.006

Appleton, J. (2020, December 3). *How the urban heat island effect is harming our cities*. Retrieved from Bee Smart City: https://hub.beesmart.city/

Appolloni, L., Giretti, A., Corazza, M., & D'alessandro, D. (2020). Walkable urban environments: An ergonomic approach of evaluation. *Sustainability*, *12*(20), 1–31. doi:10.3390u12208347

Arabi, R., Shahidan, M., Kamal, M., Jaafar, M., & Rakhshandehroo, M. (2015). Mitigating Urban Heat Island Through Green Roofs. *Current World Environment*, *10*(1), 918–927. doi:10.12944/CWE.10.Special-Issue1.111

Arafa, A. E.-S., & Hirshon, J. M. (2019). Driving behavior and road traffic crashes among professional and nonprofessional drivers in South Egypt. *International Journal of Injury Control and Safety Promotion*, *26*(4), 372–378. doi:10.1080/17457300.2019.1638419 PMID:31282807

Armson, D., Rahman, M. A., & Ennos, A. R. A. (2013). Comparison of the shading effectiveness of five different street tree species in Manchester. *Arboriculture & Urban Forestry*, *39*(4), 157–164. doi:10.48044/jauf.2013.021

Ashrafi, B., Kloos, M., & Neugebauer, C. (2021). Heritage impact assessment, beyond an assessment tool: A comparative analysis of urban development impact on visual integrity in four UNESCO World Heritage properties. *Journal of Cultural Heritage*, *47*, 199–207. doi:10.1016/j.culher.2020.08.002

Asi, Y. M. (2021, July 30). *Challenges facing sustainable development goals in Arab States*. Retrieved from Arab Center Washington DC: https://arabcenterdc.org/

Attia, S., & Khalil, H. (2015). Urban Metabolism and Quality of Life in Informal Areas. In *Proceedings REAL CORP 2015: Plan Together – Right Now – Overall*, 661-674.

Baik, J., Kwak, K. H., Park, S. B., & Ryu, Y. H. (2012). Effects of building roof greening on air quality in street canyons. *Atmospheric Environment*, *61*, 48–55. doi:10.1016/j.atmosenv.2012.06.076

Bakovic, M., Siddiqui, F., Başol, M. A., & Göçer, Ö. (2017). Outdoor Thermal Comfort Analysis at a Sustainable University Campus. In *International Symposium to Promote Innovation & Research in Energy Efficiency-INSPIRE Symposium* (*Vol. 27*) Academic Press.

Balany, F., Ng, A., & Wong, M. (2020, December). Green Infrastructure as an Urban Heat Island Mitigation Strategy- A Review. *Water (Basel)*, *12*(3577), 3577. Advance online publication. doi:10.3390/w12123577

Barlow, J. (2014). Progress in observing and modelling the urban boundary layer. *Urban Climate*, *10*, 216–240. doi:10.1016/j.uclim.2014.03.011

Basbasa, S., Campisi, T., & Canale, A. (2020). Pedestrian level of service assessment in an area close to an under-construction metro line in Thessaloniki, Greece. *Transportation Research Procedia*, *45*, 95–102. doi:10.1016/j.trpro.2020.02.067

Battuta, I. (2009). The Travels of Ibn Battuta (S. Lee, Trans.). Cosimo Classics. (Original publication 1829)

Behrens-Abouseif, D. (1989). *Islamic Architecture in Cairo: An Introduction*. E.J. Brill.

Bek, M. A., Azmy, N., & Elkafrawy, S. (2018). The effect of unplanned growth of urban areas on heat island phenomena. *Ain Shams Engineering Journal*, *9*(4), 3169–3177. doi:10.1016/j.asej.2017.11.001

Ben-Bassat, T. (2019). Are ergonomically designed road signs more easily learned? *Applied Ergonomics*, *78*, 137–147. doi:10.1016/j.apergo.2019.02.009 PMID:31046945

Ben-Bassat, T., Shinar, D., Caird, J. K., Dewar, R. E., Lehtonen, E. S., Zakowska, L., ... Pronin, M. (2021). Ergonomic design improves cross-cultural road sign comprehension. *Transportation Research Part F: Traffic Psychology and Behaviour*, *78*, 267–279. doi:10.1016/j.trf.2021.01.015

Bernatsky, A. (1978). *Tree Ecology and Preservation*. Elsevier Scientific Publishers.

Bertolini, L. (2008). Station areas as nodes and places in urban networks: An analytical tool and alternative development strategies. In Railway Development (pp. 35–57). Physica-Verlag HD. doi:10.1007/978-3-7908-1972-4_3

Bhalla, K., & Gleason, K. (2020). Effects of vehicle safety design on road traffic deaths, injuries, and public health burden in the Latin American region: A modelling study. *The Lancet. Global Health*, *8*(6), 1–10. doi:10.1016/S2214-109X(20)30102-9 PMID:32446347

Bhan, G., Srinivas, S., & Watson, V. (Eds.). (2018). *The Routledge companion to planning in the Global South*. Routledge.

Blackman, D.J. & Lentini, M.C. (2010). *Ricoveri per navi militari nei porti del Mediterraneo antico e medievale: atti del Workshop*. Edipuglia srl.

Blocken, B., Janssen, W. D., & van Hooff, T. (2012). CFD simulation for pedestrian wind comfort and wind safety in urban areas: General decision framework and case study for the Eindhoven University campus. *Environmental Modelling & Software*, *30*, 15–34. doi:10.1016/j.envsoft.2011.11.009

Boarnet, M., & Crane, R. (2001). The influence of land use on travel behavior: Specification and estimation strategies. *Transportation Research Part A, Policy and Practice*, *35*(9), 823–845. doi:10.1016/S0965-8564(00)00019-7

Bobkova, E., Berghauser, P. M., & Marcus, L. (2021). Towards analytical typologies of plot systems: Quantitative profile of five European cities. *Environment and Planning. B, Urban Analytics and City Science*, *48*(4), 604–620. doi:10.1177/2399808319880902

Bourbia, F., & Awbi, H. (2004). Building cluster and shading in urban canyon for hot dry climate: Part 1: Air and surface temperature measurements. *Renewable Energy*, *29*(2), 249–262. doi:10.1016/S0960-1481(03)00170-8

Bovill, C. (2015). *Sustainability in architecture and urban design*. Routledge.

Broadbent, A. M., Coutts, A., Tapper, N. J., & Demuzere, M. (2018). The cooling effect of irrigation on urban microclimate during heatwave conditions. *Urban Climate*, *23*, 309–329. doi:10.1016/j.uclim.2017.05.002

Burri, S., Frey, C., Parlow, E., & Vogt, R. (2009). CO2 fluxes and concentrations over an urban surface in Cairo/Egypt. *In Proceedings of The Seventh International Conference on Urban Climate*. Yokohama, Japan.

Cairo Governorate Metro. (2021). Retrieved 2020, from Cairo Metro Website: https://cairometro.gov.eg/en/about/1

Cairo Population 2022 (Demographics, Maps, Graphs). (n.d.). Available at: https://worldpopulationreview.com/world-cities/cairo-population

Cambridge. (2021). Retrieved 2022, from Cambridge.com: https://dictionary.cambridge.org/dictionary/english/regenerative#:~:text=regenerative%20adjective%20(GROWING)&text=relating%20to%20something%20growing%20or,powers%20of%20fire%20on%20cropland

Campbell, B., Hansen, J., Rioux, J., M Stirling, C., Twomlow, S., & Wollenberg, E. (2018). Urgent action to combat climate change and its impacts (SDG 13): Transforming agriculture and food systems. *Current Opinion in Environmental Sustainability*, *34*, 13–20. doi:10.1016/j.cosust.2018.06.005

Campos, I., Vizinho, A., Coelho, C., Alves, F., Truninger, M., Pereira, C., ... Lopes, G. P. (2016). Participation, scenarios and pathways in long-term planning for climate change adaptation. *Planning Theory & Practice*, *17*(4), 537–556. doi:10.1080/14649357.2016.1215511

CAPMAS. (2017). *Monitoring and follow up SDGs in Egypt*. CAPMAS.

Carmin, J., Anguelovski, I., & Roberts, D. (2012). Urban Climate Adaptation in the global South Planning in an Emerging Policy Domain. *Journal of Planning Education and Research*, *31*(1), 18–32. doi:10.1177/0739456X11430951

Carmona, M. (2021). *Public Places-Urban Spaces: The Dimensions of Urban Design* (3rd ed.). Routledge. doi:10.4324/9781315158457

Carson, R. (1962). *Silent spring*. Houghton Mifflin.

Carter, J. G., Cavan, G., Connelly, A., Guy, S., Handley, J., & Kazmierczak, A. (2015). Climate change and the city: Building capacity for urban adaptation. *Progress in Planning*, *95*, 1–66. doi:10.1016/j.progress.2013.08.001

Center for Sustainable Development (CSD). (2013). *Egypt's Strategy for Dealing with Slums*. American University in Cairo.

Chaline, C., & Coccossis, H. (2004). *Guidelines for urban regeneration in the Mediterranean region. UNEP, Priority Actions Programma*. Regional Activity Centre.

Chaoui, K. (2009). *Competitive cities and climate change*. OECD Publishing.

Chapple, K. (2015). *Planning sustainable cities and regions: Towards more equitable development*. Routledge.

Chatterji, T. (2020). *SDG 11 Sustainable Cities and Communities: SDG 11 and the New Urban Agenda: Global Sustainability Frameworks for Local Action*. Springer Nature. doi:10.1007/978-981-32-9927-6_12

Chebanyuk, K., Prasolenko, O., Burko, D., Galkin, A., Lobashov, O., Shevchenko, A., Usami, D. S., & Persia, L. (2020). Pedestrians influence on the traffic flow parameters and road safety indicators at the pedestrian crossing. *Transportation Research Procedia*, *45*, 858–865. doi:10.1016/j.trpro.2020.02.083

Chellamani, P., Singh, C. P., & Panigrahy, S. (2014). Assessment of the health status of Indian mangrove ecosystems using multi temporal remote sensing data. *Tropical Ecology*, *55*, 245–253.

Chifari, R., Renner, A., Lo Piano, S., Ripa, M., Bukkens, S. G. F., & Giampietro, M. (2017). Development of a municipal solid waste management decision support tool for Naples, Italy. *Journal of Cleaner Production*, *161*, 1032–1043. doi:10.1016/j.jclepro.2017.06.074

Ciudin, R., Isarie, C., Cioca, L., Petrescu, V., Nederita, V., & Ranieri, E. (2014). Vacuum waste collection system for an historical city centre. Academic Press.

Clarke, T., McNamara, K. E., Clissold, R., & Nunn, P. D. (2019). Community-based adaptation to climate change: Lessons from Tanna Island, Vanuatu. *Island Studies Journal*, *14*(1), 59–80. doi:10.24043/isj.80

Climate Urban Lab. (2016). *The benefits of green infrastructure for heat mitigation and emissions reductions in cities*. The Trust for Public Land.

Costa, A. P., Campos, F., & Villarouco, V. (2012). Overview of ergonomics built environment. *Work (Reading, Mass.)*, *41*, 4142–4148. doi:10.3233/WOR-2012-0710-4142 PMID:22317357

Cottrill, C. D., & Derrible, S. (2015). Leveraging big data for the development of transport sustainability indicators. *Journal of Urban Technology*, *22*(1), 45–64. doi:10.1080/10630732.2014.942094

Craft, W., Ding, L., & Prasad, D. (2021). Developing a decision-making framework for regenerative precinct development. *Sustainability*, *13*(22), 12604. doi:10.3390/su132212604

Dagnachew, A., Hof, A., Soest, H. v., & Vuuren, D. v. (2021). *Climate change measures and sustainable development goals*. PBL Netherlands Environmental Assessment Agency.

Dai, X., Sun, L., & Xu, Y. (2018). Short-term origin-destination based metro flow prediction. *Journal of Advanced Transportation*, *2018*, 1–18. doi:10.1155/2018/5942763

Darinka Czischke, C. M. (2015). *Sustainable regeneration in urban areas*. URBACT.

Das, J. (2019). *Reporting climate change in the Global North and South: Journalism in Australia and Bangladesh*. Routledge. doi:10.4324/9780429402210

David, K., & Bohl, T. L. (2018). Sustainable Development Goals Report: Egypt 2030. United Nations.

De Carvalho, R. (2012). ERGOPOLIS: An ergonomics approach applied to a city. *Work (Reading, Mass.)*, *41*, 6071–6078. doi:10.3233/WOR-2012-1063-6071

Demuzere, M., Orru, K., Heidrich, O., Olazabal, E., Geneletti, D., Orru, H., Bhave, A. G., Mittal, N., Feliu, E., & Faehnle, M. (2014). Mitigating and Adapting to Climate Change: Multi-functional and Multiscale assessment of Green Urban Infrastructure. *Journal of Environmental Management*, *146*, 107–115. doi:10.1016/j.jenvman.2014.07.025 PMID:25163601

Department of Environment, Land, Water and Planning. (2017). *Planning a Blue-Green City: A how-to guide for planning urban greening and enhanced storm water management in Victoria*. E2design Lab.

Devisscher, T., Konijnendijk, C., Nesbitt, L., Lenhart, J., Salbitano, F., Cheng, Z. C., & Bosch, M. v. (2019). SDG 11: Sustainable cities and communities – Impacts on forests and forest-based livelihoods. In P. Katila & C. J. Colfer (Eds.), *Sustainable development goals: Their impacts on forests and people* (pp. 349–385). Cambridge University Press. doi:10.1017/9781108765015.013

Dhar, T. K., & Khirfan, L. (2016). Community-based adaptation through ecological design: Lessons from Negril, Jamaica. *Journal of Urban Design*, *21*(2), 234–255. doi:10.1080/13574809.2015.1133224

Dimoudi, A., Kantzioura, A., Zoras, S., Pallas, C., & Kosmopoulos, P. (2013). Investigation of urban microclimate parameters in an urban center. *Energy and Building*, *64*, 1–9. doi:10.1016/j.enbuild.2013.04.014

DIY vine arch. (n.d.). Retrieved February 25, 2022, from https://www.cerceis.com/1432/

DIY vine sheds. (2020, February 5). *DIY garden*. https://ar.imadeself.com/ yagody/ karkas-dlya-vinograda.html

Dobrowolska, A., & Dobrowolsky, J. (2007). *Heliopolis rebirth of the city of the sun*. American University in Cairo Press.

Dohrmann Consulting. (2014, July 5). *What is Ergonomics?* Retrieved January 2022, from Dohrmann Consulting: https://www.ergonomics.com.au/about-us/

Dovey, K., & Pafka, E. (2017). What is functional mix? An assemblage approach. *Planning Theory & Practice*, *18*(2), 24–9267. doi:10.1080/14649357.2017.1281996

Dul, J., Bruder, R., Buckle, P., Carayon, P., Falzon, P., Marras, W. S., Wilson, J. R., & van der Doelen, B. (2012). A strategy for human factors/ergonomics: Developing the discipline and profession. *Ergonomics*, *55*(4), 377–395. doi:10.1080/00140139.2012.661087 PMID:22332611

Du, Y., Mak, C. M., Liu, J., Xia, Q., Niu, J., & Kwok, K. C. (2017). Effects of lift-up design on pedestrian level wind comfort in different building configurations under three wind directions. *Building and Environment*, *117*, 84–99. doi:10.1016/j.buildenv.2017.03.001

Dwyer, J. F., Schroeder, H. W., & Gobster, P. H. (1994). The Deep Significance of Urban Trees and Forests. In R. H. Platt, R. A. Rowntree, & P. C. Muick (Eds.), *The Ecological City: Preserving and Restoring Urban Biodiversity*. University of Massachusetts Press.

Egyptian Streets. (2020, October 26). *Egypt ranks first regionally for sustainable development goals assessment*. Retrieved from Egyptian Streets: https://egyptianstreets.com/

Eilouti, B. (2009). A digital incorporation of ergonomics into architectural design. *International Journal of Architectural Computing*, *7*(2), 235–253. doi:10.1260/147807709788921994

Eilouti, B. (2021). A framework for integrating ergonomics into architectural design. *Ergonomics in Design*, 1–9. doi:10.1177/1064804620983672

Eissa, Y., & Khalil, H. A. (2021). Urban Climate Change Governance within Centralised Governments: A Case Study of Giza, Egypt. *Urban Forum*. doi:10.1007/s12132-021-09441-9

El Bardisy, N., Reuter, W., & Ismail, A. (2019). Social interaction dynamics exploration: The case Study of el sherouk city. *International Journal of Architectural and Environmental Engineering*, *13*(11), 658–667.

El Darwish, I., Ragheb, A., & Sherif, A. (2015). The Role of Simulation in Urban Design Decisions: Microclimate and Human Comfort Considerations in Planning. *Architecture and Planning Journal, 23*(1).

Elbardisy, W. M., Salheen, M. A., & Fahmy, M. (2021). Solar irradiance reduction using optimized green infrastructure in arid hot regions: A case study in El-Nozha District, Cairo, Egypt. *Sustainability*, *13*(14), 9617. Advance online publication. doi:10.3390u13179617

Eldar, R., & Fisher-Gewirtzman, D. (2019). Ergonomic design visualization mapping-developing an assistive model for design activities. *International Journal of Industrial Ergonomics*, *74*, 102859. Advance online publication. doi:10.1016/j.ergon.2019.102859

Elias, A. (2011). Automobile-oriented or complete street? Pedestrian and bicycle level of service in the new multimodal paradigm. *Transportation Research Record: Journal of the Transportation Research Board*, *2257*(1), 80–86. doi:10.3141/2257-09

Ellen MacArthur Foundation. (2017). *Cities in the Circular Economy: An Initial Exploration.* Available online:https://www.ellenmacarthurfoundation.org/publications/cities-in-the-circular-economy-an-initial-exploration

El-magd, I.A., Ismail, A. & Zanaty, N. (2016). Spatial variability of urban heat islands in Cairo City, Egypt using time series of Landsat Satellite images. *International Journal of Advanced Remote Sensing and GIS*, *5*, 1618.

El-Magd, I., Ismail, A., & Zanaty, N. (2016). Spatial variability of urban heat islands in Cairo City, Egypt using time series of Landsat satellite images. *International Journal of Advanced Remote Sensing and GIS*, *5*(3), 1618–1638. doi:10.23953/cloud.ijarsg.48

Elmarakby, E., Khalifa, M., Elshater, A., & Afifi, S. (2022). Tailored methods for mapping urban heat islands in Greater Cairo Region. *Ain Shams Engineering Journal*, *13*(2), 101545. doi:10.1016/j.asej.2021.06.030

ElMoghazi, Y. (2019). *Road safety challenges in Egypt: A discussion of policy alternatives.* Retrieved 2022, from Theses and Dissertations: https://fount.aucegypt.edu/etds/790/

Elnabawi, M. H., & Hamza, N. (2020). Behavioural perspectives of outdoor thermal comfort in urban areas: A critical review. *Atmosphere*, *11*(1), 1–23. doi:10.3390/atmos11010051 PMID:32461808

Elshater, A. (2019). The Predicament of Post-Displacement Amidst Historical sites: A design-based correlation between people and place. *Heritage & Society*, *12*(2-3), 85–115. doi:10.1080/2159032X.2021.1879355

Elshater, A., Abusaada, H., & Afifi, S. (2019). What makes livable cities of today alike? Revisiting the criterion of singularity through two case studies. *Cities (London, England)*, *92*, 273–291. doi:10.1016/j.cities.2019.04.008

Elshater, A., Abusaada, H., Alfiky, A., El-Bardisy, N., Elmarakby, E., & Grant, S. (2022). Workers' satisfaction vis-à-vis environmental and socio-morphological aspects for sustainability and decent work. *Sustainability*, *14*(3), 1–34. doi:10.3390u14031699

Elton, H. A. L. A., & Shkodrani, N. (2020). Design for wind comfort. The CFD assessment over a future outdoor public space. *Academic Platform Journal of Natural Hazards and Disaster Management*, *1*(1), 13–24.

ENVI-Met. (2021, October 21). *ENVI-MET: A holistic microclimate model.* Retrieved December 25, 2021, from Envi-Met: https://envi-met.info/doku.php?id=root:start

Ermida, S. L., Soares, P., Mantas, V., Göttsche, F. M., & Trigo, I. F. (2020). Google Earth Engine open-source code for Land Surface Temperature estimation from the Landsat series. *Remote Sensing*, *12*(9), 1471.

Eurostat. (2020, May). *SDG 11 - Sustainable cities and communities.* Retrieved from Eurostat Statistics Explained: https://ec.europa.eu/

Fadl, M. S., & Karadelis, J. (2013). CFD simulation for wind comfort and safety in urban area: A case study of Coventry University central campus. *International Journal of Architecture, Engineering and Construction, 2*(2), 131-143.

Fahed, J., Ginestet, S., & Adolphe, L. (2020). Impact of urban heat island mitigation measures on microclimate and pedestrian comfort in a dense urban district of Lebanon. *Sustainable Cities and Society*, *61*, 1–14. doi:10.1016/j.scs.2020.102375

Fahmi, W., & Sutton, K. (2010). Cairo's Contested Garbage: Sustainable Solid Waste Management and the Zabaleen's Right to the City. *Sustainability*, *2*(6), 1765–1783. doi:10.3390u2061765

Falvo, M. C., Lamedica, R., Bartoni, R., & Maranzano, G. (2011). Energy management in metro-transit systems: An innovative proposal toward. *Electric Power Systems Research*, *81*(12), 2127–2138. doi:10.1016/j.epsr.2011.08.004

Fang, Z., Tang, T., Zheng, Z., Zhou, X., Liu, W., & Zhang, Y. (2021). Thermal responses of workers during summer: An outdoor investigation of construction sites in South China. *Sustainable Cities and Society*, *66*, 1–14. doi:10.1016/j.scs.2020.102705

Faragallah, R. N., & Ragheb, R. (2022). Evaluation of thermal comfort and urban heat island through cool paving materials using ENVI-Met. *Ain Shams Engineering Journal*, *13*(3), 1–13. doi:10.1016/j.asej.2021.10.004

Farahat, B. I. (2020). Employing sustainable development concepts to revitalize the historic urban quarter of Al-Muizz Street, Cairo, Egypt. Academic Press.

Fares, A. (Ed.). (2021). *Climate change and extreme events*. Elsevier Science Publishing Co Inc.

Farshid Aram, E. H. (2019). Urban green space cooling effect in cities. *Heliyon*. PMID:31008380

Fayed, L., Elshater, A., & Rashed, R. (2020). Aspects of regenerative cities. In S. Kamel, H. Sabry, G. F. Hassan, M. Refat, A. Elshater, A. S. Elrahman, . . . R. Rashed (Eds.), Architecture and Urbanism: A Smart Outlook. doi:10.1007/978-3-030-52584-2_22

Feather, J. (2006). Managing the documentary heritage: issues for the present and future. In G. Gorman (Ed.), *Preservation management for libraries, archives and museums* (pp. 1–18). doi:10.29085/9781856049214.002

Fernando, S., Fernando, S., & Mendis, P. (2018, December). Pedestrian wind comfort study using computational fluid dynamic (CFD) simulation. In *International Conference on Sustainable Built Environment* (pp. 323-339). Springer.

Fibiger, T. (2015). Heritage erasure and heritage transformation: How heritage is created by destruction in Bahrain. *International Journal of Heritage Studies*, *21*(4), 390–404. doi:10.1080/13527258.2014.930064

Filho, W. L., Wolf, F., Castro-Díaz, R., Li, C., Ojeh, V. N., Gutiérrez, N., ... Böneck, J. (2021). Addressing the urban heat islands effect: A cross-country assessment of the role of green infrastructure. *Sustainability*.

Filho, W., Wolf, F., & Castro-Diaz, R. (2020, January). Addressing the Urban Heat Island Effect: A Cross-Country Assessment of the Role of Green Infrastructure. *Sustainability*, *13*(753), 753. Advance online publication. doi:10.3390u13020753

Foggin, J. M., Brombal, D., & Razmkhah, A. (2021). Thinking like a mountain: Exploring the potential of relational approaches for transformative nature conservation. *Sustainability*, *13*(22), 12884. doi:10.3390u132212884

Forino, G., Meding, J., & Brewer, G. J. (2015). A conceptual governance framework for climate change adaptation and disaster risk reduction integration. *International Journal of Disaster Risk Science*, *6*(4), 372–384. doi:10.100713753-015-0076-z

Frantzeskaki, N., McPhearson, T., Collier, M. J., Kendal, D., Bulkeley, H., Dumitru, A., Walsh, C., Noble, K., van Wyk, E., Ordóñez, C., Oke, C., & Pintér, L. (2019). Nature-based solutions for urban climate change adaptation: Linking science, policy, and practice communities for evidence-based decision-making. *Bioscience*, *69*(6), 455–466. doi:10.1093/biosci/biz042

Fry, T. (2015). *City futures in the age of a changing climate*. Routledge.

Fusaro, G., & Kang, J. (2021). Participatory approach to draw ergonomic criteria for window design. *International Journal of Industrial Ergonomics*, *82*, 103098. Advance online publication. doi:10.1016/j.ergon.2021.103098

Gaber, D. A. (2019). *From Waste Management to Waste Markets: Infrastructure*. Industry and Business Opportunities.

Gartland, L. (2008). *Heat Islands: Understanding and Mitigating Heat in Urban Areas*. London Earthscan publishing.

Gaudet, C. (1985). Sunspots, landscaping for energy efficiency. *Harrowsmith*, *1*(61), 20–33.

Gehl, J., & Rogers, L. R. (2010). *Cities for People*. Island Press.

General Organization of Physical Planning GOPP. (2009). *Cairo Future Vision 2050*. Author.

General Organization of Physical Planning GOPP. (2012). *Greater Cairo Urban Development Strategy*. Author.

*General Organization of Physical Planning GOPP*. (2014). Ministry of Housing, Utilities and Urban Communities.

Gessa, A., & Sancha, P. (2020). Environmental open data in urban platforms: An approach to the big data life cycle. *Journal of Urban Technology*, *27*(1), 27–45. doi:10.1080/10630732.2019.1656934

Ghiaus, C., Allard, F., Santamouris, M., Georgakis, C., & Nicol, F. (2006). Urban environment influence on natural ventilation potential. *Building and Environment*, *41*(4), 395–406.

Ghoneim, E. (2009). A remote sensing study of some impacts of global warming on the Arab region. In M. K. Tolba & N. Saab (Eds.), *Arab Environment: Impact of Climate Change on the Arab Countries* (pp. 31–46). Arab Forum for Environment and Development.

Gidarakos, E., Havas, G., & Ntzamilis, P. (2006). Municipal solid waste composition determination supporting the integrated solid waste management system in the island of Crete. *Waste Management (New York, N.Y.)*, *26*(6), 668–679. doi:10.1016/j.wasman.2005.07.018 PMID:16207528

Gill, S., Handley, J., Ennos, A., & Pauleit, S. (2007). Adapting cities for climate change: The role of the green infrastructure. *Built Environment*, *33*(1), 115–133. doi:10.2148/benv.33.1.115

Girardet, H. (2014). *Creating regenerative cities*. Routledge. doi:10.4324/9781315764375

Givoni, B. (1998). *Climate Considerations in Building and Urban Design*. John Wiley & Sons.

Global Designing Cities Initiative. (2016). *Global Street Design Guide*. Island Press.

Gomez, F., Tamarit, N., & Jaboloyes, J. (2001). Green zones, bio-climatic studies and human comfort in the future development of urban planning. *Landscape and Urban Planning*, *55*, 151–161.

Governorate, G. (2018). *Towards a Climate Change Strategy in Giza Governorate Framework Document*. Giza Governorate. Retrieved from https://egwuf.files.wordpress.com/2020/01/giza-climate-change-strategy.pdf

Grig & Chris. (2011). *The New Green Spaces in Cities-The Green Roofs*. http://www.howtobuildahouseblog.com/new-green-spaces-in-cities-green-roofs

Griloa, F., Pinhoa, P., Aleixoa, C., Catita, C., Silva, P., Lopes, N., ... Branquinho, C. (2020). Using green to cool the grey: Modelling the cooling effect of green spaces with a high spatial resolution. *The Science of the Total Environment*, *724*, 138182. Advance online publication. doi:10.1016/j.scitotenv.2020.138182 PMID:32408445

Guhathakurta, S., & Gober, P. (2007). The impact of the phoenix urban heat island on residential water use. *Journal of the American Planning Association*, *73*, 317–329.

Hand, G. H., Weber, R., & Bluestone, N. (2017). Regenerative cities: Moving beyond sustainability–a Los Angeles case study. *Journal of Urban Design and Mental Health*, *3*(4).

Hashem, I. A., Chang, V., Anuar, N. B., Adewole, K., Yaqoob, I., Gani, A., ... Chirmona, H. (2016). The role of big data in smart city. *International Journal of Information Management*, *35*(5), 748–758. doi:10.1016/j.ijinfomgt.2016.05.002

Hassan, M., & Amr, N. (2018). *Title: Palm Trees Festival First Round*. https://www.google.com/url?sa =i&url =https% 3A% 2F% 2Fb mitp global network. org%2 Fwp- content% 2F uploads %2F 2021% 2F04% 2Fpalm- trees- festival- 2018. pdf &psig = AOvVaw 3fOOr0 FmEwST_iEUA XgYo8 &ust = 164 590 722 329 000 0&source = images &cd = vfe & ved = 2ah UKE wjA gaW s2J v2A hVB s6Q KHW 3uC QIQ 3Yk Beg QIA BAL

Hassan, D. M., Moustafa, Y. M., & El-Fiki, S. M. (2019). Ground-floor façade design and staying activity patterns on the sidewalk: A case study in the Korba area of Heliopolis, Cairo, Egypt. *Ain Shams Engineering Journal*, *10*(3), 453–461. doi:10.1016/j.asej 2018.12.006

Hassan, G., Rashed, R., & El Nagara, S. (2022). Regenerative urban heritage model: Scoping review of paradigms' progression. *Ain Shams Engineering Journal*, 1–8. doi:10.1016/j.asej.2021.101652

Healthcare, I., & Karwowski, W. (2006). *International Encyclopedia of Ergonomics and Human Factors - 3 Volume Set* (I. Healthcare & W. Karwowski, Eds.). Taylor & Francis Group. doi:10.1201/9780849375477

Hendel, M., Azos-Diaz, K., & Tremeac, B. (2017). Behavioral adaptation to heat-related health risks in cities. *Energy and Building*, *152*, 823–829. doi:10.1016/j.enbuild.2016.11.063

Herbert Girardet/World Future Council. (2010). *Regenerative Cities*. Hafen University.

Hernandez, S., & Monzon, A. (2016). Key factors for defining an efficient urban transport interchange: Users' perception. *Cities (London, England)*, *50*, 158–167. doi:10.1016/j.cities.2015.09.009

Hideki Takebayashi, M. M. (2020). Background and purpose. In *Adaptation Measures for Urban Heat Islands*. doi:10.1016/B978-0-12-817624-5.00001-4

Hiemstra, J. A., Saaroni, H., & Amorim, J. H. (2017). The Urban Heat Island: Thermal Comfort and the Role of Urban Greening. In D. Pearlmutter & C. Calfapietra (Eds.), *The Urban Forest* (pp. 7–19). Springer. doi:10.1007/978-3-319-50280-9_2

Hirano, Y., & Fujita, T. (2012). Evaluation of the impact of the urban heat island on residential and commercial energy consumption in Tokyo. *Energy*, *37*, 371–383.

Hongbo, W. (2015). *Global sustainable development report*. United Nations.

Hornsby, C., Ripa, M., Vassillo, C., & Ulgiati, S. (2017). A roadmap towards integrated assessment and participatory strategies in support of decision-making processes. The case of urban waste management. *Journal of Cleaner Production*, *142*, 157–172. doi:10.1016/j.jclepro.2016.06.189

Hosseini, M., Javanroodi, K., & Nik, V. M. (2022). High-resolution impact assessment of climate change on building energy performance considering extreme weather events and microclimate – Investigating variations in indoor thermal comfort and degree-days. *Sustainable Cities and Society, 78*, 103634. . doi:10.1016/j.scs.2021.103634

Howeidy, A. (2009). *Cairo's Informal Areas: Between Urban Challenges and Hidden Potentials*. GIZ Project.

Huang-Lachmann, J.-T., & Lovett, C. (2016). How cities prepare for climate change: Comparing Hamburg and Rotterdam. *Cities (London, England)*, *45*, 36–44. doi:10.1016/j.cities.2015.11.001

Hutton, B. (2013). *Planning Sustainable Transport*. Routledge. doi:10.4324/9780203436066

Ilbert, R. (1985). Heliopolis: Colonial enterprise and town planning success? In A. Evin (Ed.), *The Expanding Metropolis: Coping with the Urban Growth of Cairo* (pp. 11–15). Concept Media/ Aga Khan Award for Architecture.

International Council on Monuments and Sites (ICOMOS). (2011). *ICOMOS World Report 2002-2003 on monuments and sites in danger.* Retrieved February 2022, from https://www.icomos.org/en/what-we-do/risk-preparedness/heritage-at-risk-reports/116-english-categories/resources/publications/210-icomos-world-report-2002-2003-on-monuments-and-sites-in-danger

International Labour Organization. (2021). *Principles and guidelines for human factors/ ergonomics (HFE) design and management of work systems.* Retrieved 2022, from http://www.ilo.org/global/topics/safety-and-health-at-work/news/WCMS_826603/lang--en/index.htm

Irmak, M. A., Yilmaz, S., & Dursun, D. (2017). Effect of different pavements on human thermal comfort conditions. *Atmosfera, 30*(4), 1–12. doi:10.20937/ATM.2017.30.04.06

Istanbul Bilgi University. (n.d.a). *About santralistanbul.* Retrieved November 27, 2021, from https://www.santralistanbul.org/tr/hakkinda/

Istanbul Bilgi University. (n.d.b). *About Silahtarağa Power Plant.* Retrieved November 27, 2021, from https://www.santralistanbul.org/en/silahtaraga-power-plant/

Istanbul Bilgi University. (n.d.c). *About İstanbul Bilgi University Santral Campus map.* Retrieved November 12, 2021, from https://www.bilgi.edu.tr/upload/santralistanbul-campus-accessible-campus-map/

Isyumov, N., & Davenport, A. G. (1975). The ground level wind environment in built-up areas. In *Proceedings of the 4th International Conference on Wind Effects on Buildings and Structures* (pp. 403-422). Academic Press.

Italian Agency for Environment Protection and Technical Services. (2007). *The Waste Report 2007.* Author.

Jacobs, J. (1961). *The death and life of great American cities.* Vintage Book.

Jacobson, M. (2012). *Stuttgart green corridors.* http://wwf.panda.org/?204461/Stuttgart

Jadhav, S., Mali, V., Ramachandran, P. K., & Rane, C. (2018). *Large Scale Modelling of Wind Comfort and Safety Using 'Pedestrian Comfort Analysis'-A Cloud-based App Crafted for Architects.* Centre for Computational Technologies.

Jamei, E., Rajagopalan, P., & Mahmoudian, M. (2016). Review on the impact of urban geometry and pedestrian level greening on outdoor thermal comfort. *Renewable & Sustainable Energy Reviews, 54*, 1002–1017. doi:10.1016/j.rser.2015.10.104

James, P. (2015). *Urban sustainability in theory and practice: Circles of sustainability.* Routledge.

Janssen, W. D., Blocken, B., & van Hooff, T. (2013-a). Pedestrian wind comfort around buildings: Comparison of wind comfort criteria based on whole-flow field data for a complex case study. *Building and Environment*, *59*, 547–562. doi:10.1016/j.buildenv.2012.10.012

Janssen, W., Blocken, B., & van Hooff, T. (2013-b). Use of CFD simulations to improve the pedestrian wind comfort around a high-rise building in a complex urban area. *13th Conference of International Building Performance Simulation Association*.

Jastrzebowski, W. (1857). *Outline of Ergonomics, or the science of Work Based Upon Truths Drawn from Science of Nature 1857*. Central Institute for Labour Protection.

Jauregui, E. (1993). *Bibliography of urban climate in tropical/sub-tropical areas*. Geneva WMO Publisher.

Jenks, M., & Dempsey, N. (2005). *Future Forms and Design For Sustainable Cities*. Routledge.

Jim, C. Y. (2015). Thermal performance of climber green walls: Effects of solar irradiance and orientation. *Applied Energy*, *154*, 631–643.

Johnston, J., & Newton, J. (1996). *Building Green, A Guide for Using Plants on Roofs, Walls and Pavements*. The London Ecology Unit.

Kabisch, N., Korn, H., Stadler, J., & Bonn, A. (2017). *Nature-based solutions to climate change in urban areas: Links between science, policy, and practice*. Springer. doi:10.1007/978-3-319-56091-5

Kajabad, E. N., & Ivanov, S. V. (2019). People Detection and Finding Attractive Areas by the use of Movement Detection Analysis and Deep Learning Approach. *Procedia Computer Science*, *156*, 327–337. doi:10.1016/j.procs.2019.08.209

Kapp, R. (2021). *Stuttgart: Combating the heat island effect and poor air quality with ventilation corridors and green-blue infrastructure*. Climate ADAPI. www.climate-adapt.eea.europa.eu

Kaza, S., Yao, L. C., Bhada-Tata, P., & Van Woerden, F. (2018). *What a Waste 2.0: A Global Snapshot of Solid Waste Management to 2050*. World Bank. doi:10.1596/978-1-4648-1329-0

Khalil, H. A., & Gammaz, S. (2020). Supporting informal areas resilience: reinforcing hidden green potential for a better quality of life. In J. Moore, S. Attia, A. Abdel-Kader, & A. Narasimhan (Eds.), Ecocities Now: Building the Bridge to Socially Just and Ecologically Sustainable Cities (pp. 11-40). Springer.

Khalil, H., Ibrahim, A., ElGendy, N., & Makhlouf, N. (2018). Could/should improving the urban climate in informal areas of fast-growing cities be an integral part of upgrading processes? case study, Cairo. *Urban Climate*, *24*, 63–79.

Khalil, H., & Khalil, E. E. (2015). *Energy Efficiency in The Urban Environment*. CRC press, Taylor & Francis.

Khan, S. (2015). Revitalizing historic areas; Lessons from the renovation of Saidpur Village, Islamabad. *Journal of Research in Architecture and Planning*, *18*(1), 11–22. doi:10.53700/jrap1812015_2

Khosravi, S., Haghshenas, H., & Saleh, V. (2020). Macro-scale evaluation of urban transportation demand management policies in CBD by using system dynamics case study: Isfahan CBD. *Ransportation Research Procedia*, 2671–2689. doi:10.1016/j.trpro.2020.08.246

Kim, D. (2020). Pedestrian and Bicycle Volume Data Collection Using Drone Technology. *Journal of Urban Technology*, *27*(2), 45–60. doi:10.1080/10630732.2020.1715158

Kirk, T. (2006). Framing St. Peter's: Urban planning in Fascist Rome. *The Art Bulletin*, *88*(4), 756–776. doi:10.1080/00043079.2006.10786317

Kjelgren, R., & Montague, T. (1998). Urban tree transpiration over turf and asphalt surfaces. *Atmospheric Environment*, *32*(1), 35–41.

Kleerekoper, L. (2009). *Urban Heat: Design Principles for Urban Heat Management in the Netherlands* [Master Thesis]. Faculty of Architecture. Delft University of Technology.

Kottek, M., Grieser, J., Beck, C., Rudolf, B., & Rubel, F. (2006). World Map of the Köppen-Geiger climate classification updated. *Meteorologische Zeitschrift (Berlin)*, *15*(3), 259–263. doi:10.1127/0941-2948/2006/0130

Kropf, K. (2018). Plots, property and behaviour. *Urban Morphology: International Seminar on Urban Form, 22*(1).

Kropf, K. (2017). *The handbook of urban morphology*. John Wiley & Sons Ltd. doi:10.1002/9781118747711

Lang, T. (2005). *Insights in the British debate about urban decline and urban regeneration* (No. 32). Working Paper.

Laso, J., García-Herrero, I., Margallo, M., Bala, A., Fullana-i-Palmer, P., Irabien, A., & Aldaco, R. (2019). LCA-Based Comparison of Two Organic Fraction Municipal Solid Waste Collection Systems in Historical Centres in Spain. *Energies*, *12*(7), 1407. doi:10.3390/en12071407

Lawson, T. V. (1977). The effects of wind on people in the vicinity of buildings. Academic Press.

Lerner, J. (2014). *Urban Acupuncture*. Island Press.

Lesani, A., & Miranda-Moreno, L. (2018). Development and testing of a real-time WiFi-Bluetooth system for pedestrian network monitoring, classification, and data extrapolation. *IEEE Transactions on Intelligent Transportation Systems*, *20*(4), 1484–1496. doi:10.1109/TITS.2018.2854895

Letcher, T. (Ed.). (2021). *The impacts of climate change: A comprehensive study of physical, biophysical, social, and political issues*. Elsevier Science Publishing Co Inc. doi:10.1016/C2019-0-04893-5

Lin, Y., Sun, C. A., Lin, H., Perng, H., Chung, C., Lin, C., Lai, C.-H., Chien, W.-C., & Tsai, C. (2021). Increased risk of road traffic injuries in individuals with insomnia. *Journal of Transport & Health*, *21*, 1–5. doi:10.1016/j.jth.2021.101030

Long, L. Y. Z., & Zhang, J. (2020). Spatial and dynamic perspectives on surface urban heat island and their relationships with vegetation activity in Beijing, China, based on Moderate Resolution Imaging Spectroradiometer data. *International Journal of Remote Sensing*, *41*(3), 882–896. doi:10.1080/01431161.2019.1650985

Long, X., Bei, N., Wu, J., Li, X., Tian, F., Li, X., Zhao, S., Cao, J., Tie, X., An, Z., & … . (2018). Does afforestation deteriorate haze pollution in Beijing–Tianjin–Hebei (BTH), China? *Atmospheric Chemistry and Physics*, *18*, 10869.

Losasso, M., & D'Ambrosio, V. (2014). Progetto ambientale e riqualificazione dello spazio pubblico: Il grande progetto per il centro storico di Napoli sito Unesco. *Techne. Journal of Technology for Architecture and Environment*, *7*, 64–74.

Lovejoy, T. E., & Hannah, L. (Eds.). (2019). *Biodiversity and climate change: Transforming the biosphere*. Yale University Press. doi:10.2307/j.ctv8jnzw1

Ma. (2015). Sustainable water systems for the city of tomorrow—A conceptual framework. *Sustainability*.

Machac, J., Louda, J., & Dubova, L. (2016). Green and Blue Infrastructure: An Opportunity for Smart Cities? *Smart Cities Symposium*, Prague. 10.1109/SCSP.2016.7501030

Macrotrends. (2021). *Cairo, Egypt Metro Area Population 1950-2021*. www.macrotrends.net

Ma, G., Liu, T., & Shang, S. (2021). Improving the climate adaptability of building green retrofitting in different regions: A weight correction system for Chinese national standard. *Sustainable Cities and Society*, *69*, 102843. Advance online publication. doi:10.1016/j.scs.2021.102843

Masnavi, M. R. (2011). *Sustainable urban forms planning and design strategies: the compact city, urban sprawl and mixed-use development in theory and practice*. LAP Lambert Academic Publishing.

Mazzeo, G. (2009). Naples. *Cities (London, England)*, *26*(6), 363–376. doi:10.1016/j.cities.2009.06.001

McHarg, I. (1969). *Design with nature*. John Wiley & Sons, Inc.

McPherson, E. G., Scott, K. I., & Simpson, J. R. (1998). Estimating cost effectiveness of residential yard trees for improving air quality in Sacramento, California, using existing models. *Atmospheric Environment*, *32*(1), 75–84.

Melnikov, V., Krzhizhanovskaya, V. V., & Sloot, P. M. (2017). Models of pedestrian adaptive behaviour in hot outdoor public spaces. *Procedia Computer Science*, *108*, 185–194. doi:10.1016/j.procs.2017.05.006

Middleton, J. (2011). "I'm on autopilot, I just follow the route": Exploring the habits, routines, and decision-making practices of everyday urban mobilitie. *Environment & Planning*, *43*(12), 2857–2877. doi:10.1068/a43600

Mirzaei, P. A., & Haghighat, F. (2010). Approaches to study Urban Heat Island – Abilities and limitations. *Building and Environment*, *45*(10), 2192–2201. doi:10.1016/j.buildenv.2010.04.001

Mohajerani, A., Bakaric, J., & Jeffrey-Bailey, T. (2017). The urban heat island effect, its causes, and mitigation, with reference to the thermal properties of asphalt concrete. *Journal of Environmental Management*, *197*, 522–538. doi:10.1016/j.jenvman.2017.03.095 PMID:28412623

Moore, G. M. (2020). Connectivity of public open space: Its meaning for different functions. *Journal of Urban Design*, *26*(3), 279–295. doi:10.1080/13574809.2020.1801340

Moosmann, L., Siemons, A., Fallasch, F., Schneider, L., Urrutia, C., Wissner, N., & Oppelt, D. (2021). *The COP26 climate change conference, status of climate negotiations and issues at stake*. Luxembourg: Study for the committee on the Environment, Public Health and Food Safety, Policy Department for Economic, Scientific and Quality of Life Policies, European Food Safety.

Musco, F. (2016). *Counteracting Urban Heat Island Effects in a Global Climate Change Scenario*. doi:10.1007/978-3-319-10425-6

Musi, A (2003). Napoli, una capitale e il suo regno. *Touring, 118*, 156.

Musyawaroh, M., Pitana, T. S., & Masykuri, M., & Nandariyah. (2018). Sustainable revitalisation in Cultural Heritage Kampong Kauman surakarta supported by spatial analysis. *IOP Conference Series. Earth and Environmental Science*, *123*(1).

Naboni, E., Natanian, J., Brizzi, G., Florio, P., Chokhachian, A., Galanos, T., & Rastogi, P. (2019). A digital workflow to quantify regenerative urban design in the context of a changing climate. *Renewable and Sustainable Energy*, *113*, 1–15. doi:10.1016/j.rser.2019.109255

Naci, H., Chisholm, D., & Baker, T. (2009). Distribution of road traffic deaths by road user group: a global comparison. *Injury Prevention, 15*(1), 55–59. doi:10.1136/ip.2008.018721

Nahyan, H. M. (2017). *Planning Green-Blue Infrastructure for Urban Climate Adaptation: A Study Related to Urban Flooding Issues in Dhaka* [Master Thesis]. Land Use Planning, Wageningen University, The Netherlands.

Narumi, D. (2007). Effects of temperature change upon the regional electric power consumption. *Energy and Resources*, *28*(3), 13–19.

Nathanson, J. A. (n.d.). Solid-waste management. *Encyclopedia Britannica*. https://www.britannica.com/technology/solid-waste-management

National Centers for Environmental Information. (2021). *Global Climate Report - Annual 2021*. NOAA. Available online at https://www.ncdc.noaa.gov/sotc/global/202113

Navarro, P. J., Fernández, C., Borraz, R., & Alonso, D. (2017). A machine learning approach to pedestrian detection for autonomous vehicles using high-definition 3D range data. *Sensors (Basel)*, *17*(1), 1–20. doi:10.339017010018 PMID:28025565

Norton, B., Bosomworth, K., & Coutts, A. (2013). *Planning for a Cooler Future: Green Infrastructure to Reduce Urban heat*. Climate Adaptation for Decision-Makers.

Nurjayanti, W. (2019). Micro and mezzo space pattern in Kampung Kauman Solo. AIP Conference Proceedings, 2114(1), 40015. doi:10.1063/1.5112444

Nuruzzaman, M. (2015). Urban heat island: causes, effects and mitigation measures - a review. *International Journal of Environmental Monitoring and Analysis*.

Occupational Safety and Health Administration. (1999). *Ergonomics Program*. U.S. DEPARTMENT OF LABOR.

Occupational Safety and Health Administration. (2022). *Occupational Safety and Health Administration*. Retrieved from https://www.dol.gov/

Ogujiubm, K. K., & Ehigiamusoe, K. U. (2013). The challenges and implications of sustainable development in Africa: Policy options for Nigeria. *Journal of Economic Cooperation and Development*.

Oke, T. R. (1981). Canyon geometry and the nocturnal urban heat island: Comparison of scale model and field observations. *International Journal of Climatology*, *1*(3), 237–254. doi:10.1002/joc.3370010304

Oke, T. R. (1982). The energetic basis of the urban heat island. *Quarterly Journal of the Royal Meteorological Society*, *108*(455), 1–24.

Okumus, D. E., & Terzib, F. (2021). Evaluating the role of urban fabric on surface urban heat island: The case of Istanbul. *Sustainable Cities and Society*, *73*, 103128. Advance online publication. doi:10.1016/j.scs.2021.103128

Oliveira, S., Andrade, H., & Vaz, T. (2011, November). The cooling effect of green spaces as a contribution to the mitigation of urban heat island: A case study in Lisbon. *Building and Environment*, *46*(11), 2186–2194. doi:10.1016/j.buildenv.2011.04.034

Oliveira, V. (2020). Urban Morphology. In R. Dilworth (Ed.), *Urban Studies*. Oxford University Press. doi:10.1093/obo/9780190922481-0004

OpenFOAM. (n.d.). *About OpenFOAM*. Retrieved December 21, 2021, from https://www.openfoam.com/

Pacione, M. (2009). *Urban Geography A Global Perspective*. Routledge. doi:10.4324/9780203881927

Pao, T.-H., & Kleiner, B. H. (2021). New developments concerning the occupational safety and health act. *Managerial Law*, *43*(1/2), 138–146. doi:10.1108/03090550110770462

Parry, M. I., Canziani, O. F., Patulikof, J. P., & Hanson, C. E. (2007). *Impacts, adaptation and vulnerability. Contribution of working group 11 to the Fourth Assessment Report of the Ingovernmental panel on Climate Change.* Cambridge University Press.

Payne, E., & Gordon, B. (2019). The 20th-century architecture of Frank Lloyd Wright. Chicago: Nomination to the World Heritage List by the United States of America (2016) Revised 2019.

Perini, K., & Magliocco, A. (2014). Effects of vegetation, urban density, building height, and atmospheric conditions on local temperatures and thermal comfort. *Urban Forestry & Urban Greening*, *13*, 495–506.

Pitana. (2018). *The Architecture of Omah as an Expression of the Beauty of Javanese Culture.* Academic Press.

Plessis, C. D., & Brandon, P. (2015). An ecological worldview as basis for a regenerative sustainability paradigm for the built environment. *Journal of Cleaner Production*, *109*, 53–61. doi:10.1016/j.jclepro.2014.09.098

Plocoste, T., Jacoby-Koaly, S., Molinié, J., & Petit, R. H. (2014). Evidence of the effect of an urban heat island on air quality near a landfill. *Urban Climate*, *10*, 745–757.

Poon, S. (2020). Symbolic resistance: Tradition in batik transitions sustain beauty, cultural heritage and status in the era of modernity. *World Journal of Social Science*, *7*(2), 1–10. doi:10.5430/wjss.v7n2p1

Punkkinen, H., Merta, E., Teerioja, N., Moliis, K., & Kuvaja, E. (2012). Environmental sustainability comparison of a hypothetical pneumatic waste collection system and a door-to-door system. *Waste Management (New York, N.Y.)*, *32*(10), 1775–1781. doi:10.1016/j.wasman.2012.05.003 PMID:22647502

Purwani, O. (2018). Cosmological Layout and Hierarchy. *Colonial Narrative and Structure in Javanese Palaces, 294*. Available online at https://www.academia.edu/38186759/COSMOLOGICAL_LAYOUT_AND_HIERARCHY_Final_pdf

Puvanachandra, P., Hoe, C., El-Sayed, H. F., Saad, R., Al-Gasseer, N., Bakr, M., & Hyder, A. (2012). Road traffic injuries and data systems in Egypt: Addressing the challenges. *Traffic Injury Prevention*, *13*(1), 44–56. doi:10.1080/15389588.2011.639417 PMID:22414128

Rahaman, K. R., Ojmori, N., & Harata, N. (2005). Evaluation of the roadside walkway environment of Dhaka City. *Proceedings of the Eastern Asia Society for Transportation Studies.*

Ramyar, R., Ackerman, A., & Johnston, D. M. (2021). Adapting cities for climate change through urban green infrastructure planning. *Cities (London, England)*, *117*, 103316. Advance online publication. doi:10.1016/j.cities.2021.103316

Rauken, T., Mydske, P. K., & Winsvold, M. (2015). Mainstreaming climate change adaptation at the local level. *Local Environment: The International Journal of Justice and Sustainability*, *20*(4), 408–423. doi:10.1080/13549839.2014.880412

Raymond, A. (1993). *Le Caire*. Fayard.

Razzaghmanesh, M., Beecham, S., & Salemi, T. (2016). The role of green roofs in mitigating Urban Heat Island effects in the metropolitan area of Adelaide, South Australia. *Urban Forestry & Urban Greening*, *15*, 89–102.

Read, G., Salmon, P., Goode, N., & Lenné, M. (2018). A sociotechnical design toolkit for bridging the gap between systems-based analyses and system design. *Human Factors and Ergonomics in Manufacturing & Service Industries*, *28*(6), 327–341. doi:10.1002/hfm.20769

Regenesis Group. (2016). *Regenerative development and design: A framework for evolving sustainability*. John Wiley & Sons, Inc.

Rehan, R. (2016). Cool city as a sustainable example of heat island management: Case study of the coolest city in the world. *Housing and Building National Research Center Journal*, *12*(2), 191–204. doi:10.1016/j.hbrcj.2014.10.002

Ren, C., Ng, E. Y., & Katzschner, L. (2011). Urban climatic map studies: A review. *International Journal of Climatology*, *31*, 2213–2233.

Rijsman, L., van Ort, N., Hoogendoorn, S., Molin, E., & Teijl, T. (2019). Walking and bicycle catchment areas of tram stops: Factors and insights. In *6th International Conference on Models and Technologies for Intelligent Transportation Systems* (pp. 1–5). IEEE Xpolre. 10.1109/MTITS.2019.8883361

Robaa, S. M. (2003). Urban-Suburban/Rural Differences over Greater Cairo. *Atmosfera*, *16*(3), 157–171.

Roberts, P. (2000). The evolution, definition and purpose of urban regeneration. Urban Regeneration, 9-36.

Roesler, S., & Kobi, M. (2018). *The urban microclimate as artifact: Towards an architectural theory of thermal diversity*. Birkhauser. doi:10.1515/9783035615159

Roös, P. B. (2021). *Regenerative-adaptive design for sustainable development*. Springer. doi:10.1007/978-3-030-53234-5

Rose, C., & Boxberger, L. (1995). Ottoman Cairo. In Cairo: Living Past, Living Future. The University of Texas Centre for Middle Eastern Studies.

Roshan, G., Oji, R., & Fdez-Arróyabe, P. (2022). Future impact of cloudiness and wind changes on thermal comfort conditions in the city of Tabriz (Iran) in the frame of climate change. *Sustainable Cities and Society*, *77*, 1–15. doi:10.1016/j.scs.2021.103575

Rui, L., Buccolieri, R., Gao, Z., Gatto, E., & Ding, W. (2019). Study of the effect of green quantity and structure on thermal comfort and air quality in an urban-like residential district by ENVI-met modelling. *Building Simulation*, *12*, 183–194.

Ruskin, J. (1848). *Modern painters* (Vol. 1). Smith, Elder, and Company.

Ruskin, J. (2016). *Nature studies; selections from the writings of John Ruskin*. Wentworth Press.

Salonen, M., & Toivonen, T. (2013). Modelling travel time in urban networks: Comparable measures for private car and public transport. *Journal of Transport Geography*, *31*, 1–11. doi:10.1016/j.jtrangeo.2013.06.011

Santamouris, M. (2001). *Energy and Climate in the Urban Built Environment*. London James & James Publication.

Saptaputra, S. K., Kurniawidjaja, L. M., Susilowati, I. H., & Pratomo, H. (2021). Ergonomic sofa design to support kangaroo mother care in Indonesia. *Journal of Neonatal Nursing*, *27*(6), 471–475. doi:10.1016/j.jnn.2021.06.013

Sarkar, S. (2003). Qualitative evaluation of comfort needs in urban walkways in major activity centers. *Transportation Quarterly*, *57*, 39–59.

Sawka, M., Millward, A., Mckay, J., & Sarkovich, M. (2013, May). Growing summer energy conservation through residential tree planting. *Landscape and Urban Planning*, *13*, 1–9. doi:10.1016/j.landurbplan.2013.01.006

SDG 11: Sustainable Cities and Communities. (2021). Retrieved from SDG 11: Sustainable Cities And Communities: https://in.one.un.org/page/about-us/

Secretary-General, U. N., & Ki-moon, B. A. N. (2016). *The sustainable development goals report 2016*. United Nations.

Selim, G. (2016). Between Order and Modernity: Resurgence Planning in Revolutionary Egypt. *Journal of Urban History*, *42*(1), 180–200. doi:10.1177/0096144214566980

Serteser, N., & Karadag, I. (2018). Design for improving pedestrian wind comfort: A case study on a courtyard around a tall building. *Architectural Science Review*, *61*(6), 492–499. doi:10.1080/00038628.2018.1492899

Seto, K. C., Güneralp, B., & Hutyra, L. R. (2012). Global forecasts of urban expansion to 2030 and direct impacts on biodiversity and carbon pools. *Proceedings of the National Academy of Sciences of the United States of America*, *109*(40), 16083–16088. doi:10.1073/pnas.1211658109 PMID:22988086

Shahmohamadi (Kloss), P., Che-Ani, A. I., Maulud, K. N. A., Tawil, N. M., & Abdullah, N. A. G. (2011). The Impact of Anthropogenic Heat on Formation of Urban Heat Island and Energy Consumption Balance. *Urban Studies Research*.

Shahraiyni, H. T., Sodoudi, S., El-Zafarany, A., & Seoud, T. A. (2016). A Comprehensive Statistical Study on Daytime Surface Urban Heat Island during Summer in Urban Areas, Case Study: Cairo and Its New Towns. *Remote Sensing*, *8*(643). Advance online publication. doi:10.3390/rs8080643

Shashua-Bar, L. & Hoffman, M.E. (2000). Vegetation as a Climatic Component in the Design of an Urban Street. *Journal of Energy and Buildings*, 31.

Shearer, H., Coiacetto, E., Dodson, J., & Taygfeld, P. (2016). How the structure of the Australian housing development industry influences climate change adaptation. *Housing Studies*, *31*(7), 809–828. doi:10.1080/02673037.2016.1150430

Shishegar, N. (2014). The impacts of green areas on mitigating urban heat island effect: A review. *The International Journal of Environmental Sustainability*, *9*(1), 119–130. doi:10.18848/2325-1077/CGP/v09i01/55081

Shoubi, M. V., Barough, A. S., & Rasoulijavaheri, A. (2013). Ergonomics principles and utilizing it as a remedy for probable work related injuries in construction projects. *International Journal of Advances in Engineering and Technology*, (Mar), 232–245.

Simscale. (n.d.). *About Simscale*. Retrieved December 21, 2021, from https://www.simscale.com/

Sims, D. (2012). *Understanding Cairo: The Logic of a City Out of Control*. The American University in Cairo Press.

Simulationhub. (n.d.a). *About Simulationhub*. Retrieved December 21, 2021, from https://www.simulationhub.com/

Simulationhub. (n.d.b). *About Simulationhub credits*. Retrieved December 21, 2021, from https://www.simulationhub.com/credits

Singh, C., Soundarya, G. M., Few, R., Kuchimanchi, B., Segnon, A. C., & Morchain, D. (2021). Interrogating 'effectiveness' in climate change adaptation: 11 guiding principles for adaptation research and practice. *Climate and Development*, 1–15. Advance online publication. doi:10.1080/17565529.2021.1964937

Soares Henriques, L., John, N., Harseim, L., Ababei, P., & Mujica von Seggern, M. J. (2017). *Climate Responsive Urban Planning in El-Gamaleya*. Decentralized Greywater Recycling.

Sohn, K. (2013). Feature mapping the Seoul Metro station areas based on a self-organizing map. *Journal of Urban Technology*, *20*(4), 23–42. doi:10.1080/10630732.2013.855514

Solecki, W. D., Rosenzweig, C., Parshall, L., Pope, G., Clark, M., & Cox, J. (2005). Mitigation of the heat island effect in urban New Jersey. *Global Environmental Change Part B: Environmental Hazards*, *6*(1), 39–49.

Soltani, A., & Sharifi, E. (2019). Understanding and analysing the urban heat island (UHI) effect in Micro-Scale. *International Journal of Social Ecology and Sustainable Development*, *10*(2), 14–28. doi:10.4018/IJSESD.2019040102

Spyridi, D., Vlachokostas, C., Michailidou, A. V., Sioutas, C., & Moussiopoulos, N. (2015). Strategic planning for climate change mitigation and adaptation: The case of Greece. *International Journal of Climate Change Strategies and Management*, *7*(3), 272–289. doi:10.1108/IJCCSM-02-2014-0027

Steelyana, E. (2012). Batik, A Beautiful Cultural Heritage that Preserve Culture and Supporteconomic Development in Indonesia Authors. *BINUS Business Review*, *3*(1), 116. Advance online publication. doi:10.21512/bbr.v3i1.1288

Stevens, N. J., & Salmon, P. M. (2015). New knowledge for built environments: Exploring urban design from socio-technical system perspectives. in Artificial Intelligence and Lecture Note. In D. Harris (Ed.), *Engineering Psychology and Cognitive Ergonomics* (Vol. 9174). Springer. doi:10.1007/978-3-319-20373-7_19

Stojanovski, T. (2020). Urban design and public transportation–public spaces, visual proximity and Transit-Oriented Development (TOD). *Journal of Urban Design*, *25*(1), 134–154. doi:10.1080/13574809.2019.1592665

Sun, G., Zacharias, J., Ma, B., & Oreskovic, N. (2016). How do metro stations integrate with walking environments? Results from walking access within three types of built environment in Beijing. *Cities (London, England)*, *56*, 91–98. doi:10.1016/j.cities.2016.03.001

Swelim, T. (2015). *Ibn Tulun: His Lost City and Great Mosque*. American University in Cairo Press.

Tabari, H., & Willems, P. (2018). Seasonally varying footprint of climate change on precipitation in the Middle East. *Scientific Reports*, *8*(4435), 1–10. doi:10.103841598-018-22795-8 PMID:29535418

Takase, C. (2018). *Implementing SDG 11 – key elements, challenges and opportunities*. The United Nations Centre for Regional Development.

Takebayashi, H., & Moriyama, M. (2007). Surface heat budget on green roof and high reflection roof for mitigation of urban heat island. *Building and Environment*, *42*, 2971–2979.

Tan, N. H., Kolokotsa, D. D., & Takebayashi, H. (2021, January 26). *Greenery as a mitigation and adaptation strategy to urban heat*. Retrieved from Nature reviews earth & environment: https://www.nature.com/

Tan, J., Zheng, Y., Tang, X., Guo, C., Li, L., Song, G., Zhen, X., Yuan, D., Kalkstein, A. J., & Li, F. (2010). The urban heat island and its impact on heat waves and human health in Shanghai. *International Journal of Biometeorology*, *54*, 75–84.

Tanrıkul, A., & Hoşkara, Ş. (2019). A new framework for the regeneration process of Mediterranean historic city centres. *Sustainability*, *11*(16), 4483. doi:10.3390u11164483

Tarek, M., Hassan, G. F., Elshater, A., & Elfayoumy, M. (2021). Investigating built environment indicators to develop a local walkability index. *Journal of Contemporary Urban Affairs*, *5*(2), 235–251. doi:10.25034/ijcua.2021.v5n2-7

Tayyari, F., & Smith, J. L. (1997). *Occupational Ergonomics: Principles and application*. Kluwer Academic Publishers.

The Global Goals. (2015). *The Global Goals*. Retrieved from The Global Goals for Sustainable Development: https://www.globalgoals.org/

The International Ergonomics Association (IEA). (2022). *History of The IEA*. Retrieved from IEA: https://iea.cc/history/

The Ministry of International Cooperation. (2016). *Egypt national review report*. The Ministry of International Cooperation.

The Ministry of Planning and Economic Development. (2021). *Egypt'S 2021 voluntary national review*. The Ministry of Planning and Economic Development.

The Ministry Of Planning and Follow-up and Administrative Reform. (2016). *Sustainable development strategy (SDS): Egypt vision 2030*. The Ministry Of Planning and Follow-up and Administrative Reform.

The Ministry of Planning, Monitoring, and Administrative Reform. (2018). Egypt's national voluntary report 2018. Cairo: The Ministry of Planning, Monitoring and Administrative Reform.

The United States Environmental Protection Agency (EPA). (2008). https://www.epa.gov/heatislands

Tim, W. (1985). Heliopolis: a colonial enterprise that became an urban success. In A. Evin (Ed.), *The expanding metropolis: Coping with the urban growth of Cairo* (pp. 11–15). Concept Media/Aga Khan Award for Architecture.

Tovey, M., Woodcock, A., & Osmond, J. (2017). *Designing Mobility and Transport Services*. Taylor and Francis.

Tsichritzis, L., & Nikolopoulou, M. (2019). The effect of building height and façade area ratio on pedestrian wind comfort of London. *Journal of Wind Engineering and Industrial Aerodynamics*, *191*, 63–75. doi:10.1016/j.jweia.2019.05.021

Tsoka, S. (2017). Investigating the relationship between urban spaces morphology and local microclimate: A study for Thessaloniki. *Procedia Environmental Sciences*, *38*, 674–681.

U.S. Environmental Protection Agency. (2008). *Heat Island Effect: Cool Pavement Product Information. Environmental Protection Agency*. https://www.epa.gov/ heatisld/ images/ extra/level3_pavingproducts.html

U.S. Environmental Protection Agency. (2016). *U.S. Environmental Protection Agency*. Retrieved from Reduce Urban Heat Island Effect: https://www.epa.gov/

Ulrich, R. S., & Parsons, R. (1992). Influences of Passive Experiences with Plants on Individual Well-being and Health. In D. Relf (Ed.), *The Role of Horticulture in Human Well-being and Social Development: A National Symposium*. Timber Press Inc.

UN Climate Change News. (2021, August 9). *Secretary-General's Statement on the IPCC Working Group 1 Report on the Physical Science Basis of the Sixth Assessment*. Retrieved from United Nations Climate Change: https://unfccc.int

UNEP, UNEP DTU Partnership, & World Adaptation Science Programme (WASP). (2021). *Adaptation Gap Report 2020*. United Nations Environment Programme.

UN-Habitat. (2014). *Urban planning for city leaders* UN-Habitat.

UN-Habitat. (n.d.). *Urban Regeneration*. Retrieved from UN-Habitat: https://unhabitat.org/topic/urban-regeneration

United Nations Environment Programme. (2016). *Healthy environment, healthy people*. United Nations.

United Nations Framework Convention on Climate Change (UNFCCC). (2016). *The Paris agreement*. United Nations.

United Nations. (2015). *Sustainable Development Goals*. Retrieved from United Nations: https://www.un.org/

United Nations. (2015). *Take Action for the Sustainable Development Goals*. Retrieved from Sustainable Development Goals: https://www.un.org/

United Nations. (2016). *Sustainable Development Goals*. Retrieved from Take Action for the Sustainable Development Goals: https://www.un.org/

United Nations. (2016). The World's [Department of Economic and Social Affairs, Population Division.]. *Cities (London, England)*, •••, 2016.

United Nations. (2017). *Transforming our world: the 2030 Agenda for Sustainable Development*. United Nations.

United Nations. (2019). *The Sustainable Development Goals Report 2019*. United Nations.

Urban Climate Lab. (2016). *The benefits of green infrastructure for heat mitigation and emissions reductions in cities: A review of the literature*. The Trust for Public Land's Climate-Smart Cities Program.

Vandenbussche, L. (2018). Mapping stakeholders' relating pathways in collaborative planning processes; A longitudinal case study of an urban regeneration partnership. *Planning Theory & Practice*, *19*(4), 534–557. doi:10.1080/14649357.2018.1508737

Vanos, K. J. (2015). Children's health and vulnerability in outdoor microclimates: A comprehensive review. *Environment International*, *76*, 1–15. doi:10.1016/j.envint.2014.11.016 PMID:25497108

Voogt, G. A. (2004). *Urban Heat Islands: Hotter Cities*. Actionbioscience E-Newletter.

Wamsler, C., Wickenberg, B., Hanson, H., Olsson, J. A., Stålhammar, S., Björn, H., ... Zelmerlow, F. (2020). Environmental and climate policy integration: Targeted strategies for overcoming barriers to nature-based solutions and climate change adaptation. *Journal of Cleaner Production*, *247*, 119154. Advance online publication. doi:10.1016/j.jclepro.2019.119154

Wanderlingh, A. (2010). Naples: The History. Intra Moenia.

Wang, J., Han, S. H., & Li, X. (2021). 3D fuzzy ergonomic analysis for rapid workplace design and modification in construction. *Automation in Construction*, *123*, 1–14. doi:10.1016/j.autcon.2020.103521

Waste Management in Cairo: An Informal Success. (2019). *The Borgen Project*. Available at: https://borgenproject.org/waste-management-in-cairo/

Watkins, R., Palmer, J., & Kolokotroi, M. (2007). Increased temperature and intensification of the urban heat island: Implications for human comfort and urban design. *Built Environment*, *33*(1), 85–96. doi:10.2148/benv.33.1.85

Weihe, W. H. (1986). *Life Expectancy and its Applications with Special Regard to Tropical Areas*. World Meteorological Organization.

Westendorff, V. (2021). *Urban Trees: Strategies for Reducing Urban Heat Island in Cities*. The Field. ASLA Professional Practice Networks' blog. www.thefeild.asla.org

Wilson, J. (2014). Fundamentals of systems ergonomics/human factors. *Applied Ergonomics*, *45*(1), 5–13. doi:10.1016/j.apergo.2013.03.021 PMID:23684119

Winter, T. (2016). Climate change and our heritage of low carbon comfort. *International Journal of Heritage Studies*, *22*(5), 382–394. doi:10.1080/13527258.2016.1157092

Wong, N., & Yu, C. (2005, September). Study of green areas and urban heat island in a tropical city. *Habitat International*, *29*(3), 547–558. doi:10.1016/j.habitatint.2004.04.008

World Bank Group. (2021). *Country: Egypt. Climate Change Knowledge Portal*. https://climateknowledgeportal.worldbank.org/ country /egypt/ climate-sector-health

World Health Organization (WHO). (2012). *Health Indicators of Sustainable Cities in the Context of the Rio+20 UN Conference on Sustainable Development*. WHO.

World Health Organization. (2018). *Global status report on road safety 2018*. World Health Organization.

World Health Organization. (2021, June 21). *Road traffic injuries*. Retrieved 2022, from Newsrooms: https://www.who.int/news-room/fact-sheets/detail/road-traffic-injuries

World Health Organization. (2021, June 21). *Road traffic injuries*. Retrieved February 2022, from https://www.who.int/news-room/fact-sheets/detail/road-traffic-injuries

World Health Organization. (2022). *Estimated road traffic death rate (per 100 000 population)*. Retrieved 2022, from Rate: https://www.who.int/data/gho/indicator-metadata-registry/imr-details/198

World Health Organization. (n.d.). WHO Global Ambient Air Quality Database (2018). In *Atlas of Sustainable Development Goals 2020*. https://datatopics.worldbank.org/ sdgatlas/ goal -11- sustainable -cities -and- communities/ #footnote9

World Meteorological Organization (WMO). (2021). *State of the climate in Africa 2020*. Geneva: WMO. Retrieved from https://library.wmo.int/index.php?lvl=notice_display&id=21973#.YckILRNBz6O

World Meteorological Organization (WMO). (2021). *Urban heat island.* Retrieved from World Meteorological Organization (WMO): https://community.wmo.int/

Wouters, P., Dresiseitl, H., & Wanschura, B. (n.d.). Blue-Green Infrastructure as Tools for the Management of Urban Development and the Effects of Climate Change. *Ramboll Environ.*

Wright, F. L. (2012). *Drawings and plans of Frank Lloyd Wright: The early period (1893-1909).* Dover Publications.

Wu, Z., & Chen, L. (2017). Optimizing the spatial arrangement of trees in residential neighborhoods for better cooling effects: Integrating modeling with in-situ measurements. *Landscape and Urban Planning*, *67*, 463–472.

Yang, J., Hu, X., & Marvin, S. (2021). Verifying an ENVI-met simulation of the thermal environment of Yanzhong Square Park in Shanghai. *Urban Forestry & Urban Greening*, *66*, 1–12. doi:10.1016/j.ufug.2021.127384

Yang, L., Lam, J. C., & Tsang, C. L. (2008). Energy performance of building envelopes in different climate zones in China. *Applied Energy*, *85*, 800–817.

Yang, L., Qian, F., Song, D.-X., & Zheng, K.-J. (2016). Research on urban heat-island effect. *Procedia Engineering*, *169*, 11–18. doi:10.1016/j.proeng.2016.10.002

Yang, W., Lin, Y., & Li, C.-Q. (2018). Effects of landscape design on urban microclimate and thermal comfort in tropical climate. *Advances in Meteorology*, *2018*, 1–13. doi:10.1155/2018/2809649

Yigitcanlar, T., Kamruzzaman, M., Buys, L., Ioppolo, G., Sabatini-Marques, J., da Costa, E. M., & Yun, J. H. J. (2018). Understanding 'smart cities': Intertwining development drivers with desired outcomes in a multidimensional framework. *Cities (London, England)*, *81*, 145–160. doi:10.1016/j.cities.2018.04.003

Yigitcanlar, T., Kankanamge, N., Regona, M., Maldonado, A. R., Rowan, B., Ryu, A., ... Li, R. Y. (2020). Artificial intelligence technologies and related urban planning and development concepts: How are they perceived and utilized in Australia? *Journal of Open Innovation*, *6*(4), 1–21. doi:10.3390/joitmc6040187

Zafri, N. M., Rony, A. I., & Adri, N. (2019). Analysis of pedestrian crossing speed and waiting time at intersections in Dhaka. *Infrastructures*, *4*(3), 1–20. doi:10.3390/infrastructures4030039

Zakaria, J., & Ujang, N. (2015). Comfort of walking in the city center of Kuala Lumpur. *Procedia: Social and Behavioral Sciences*, *170*, 642–652. doi:10.1016/j.sbspro.2015.01.066

Zhang, X., Weerasuriya, A. U., Zhang, X., Tse, K. T., Lu, B., Li, C. Y., & Liu, C. H. (2020, December). Pedestrian wind comfort near a super-tall building with various configurations in an urban-like setting. *Building Simulation*, *13*(6), 1385–1408. doi:10.100712273-020-0658-6 PMID:32837689

# About the Contributors

**Hisham Abusaada** is a professor of architecture and urban design in the Housing and Building National Research Centre (HBRC), Egypt. During 1995-2015, he taught urban design course at universities in Saudi Arabia and Egypt. He is the author and co-authors of several scientific manuscripts and books in both Arabic and English published in scientific journals and conference proceedings. His area of interest is in urban design between theory and practice.

**Abeer Elshater**, PhD, is a professor of Urban Morphology at Ain Shams University (ASU), Cairo, Egypt, teaching and supervising multidisciplinary topics in the ideology of urban design. Her area of interest is in urban design disciplines. She has worked on some international research projects with international universities. Elshater is an author and co-author of seven books as well as a number of scientific manuscripts in international journals and conference proceedings. Recently, she acts as an ambassador of the Regional Studies Association, United Kingdom.

* * *

**Hala Hussien Aboulnaga** is an MSc student in the Urban Planning and Design Department at Ain Shams University, Cairo, Egypt. Hala is also a junior urban designer and architect at the AAR office. She received her BSc in urban planning and design in 2020 from Ain Shams University. She is interested in urbanism, urban planning, and sustainable landscapes, in addition to her passion for achieving greener societies.

**Samy Afifi**, PhD, is an assistant professor at the Department of Urban Planning and Design, Faculty of Engineering, Ain Shams University. He is a member of a team that is responsible for teaching many courses such as Theories and history of urban planning, urban economics, city planning and housing, and computer applications for urban planning. In Professional practice, Samy is a co-founder of Urban Development Centre. His professional career allowed him to act, discuss, negotiate and deal with a spectrum of people with different backgrounds and governmental organisations all over Egypt. His primary interest is in tracing and analyzing dynamics during the evolution of traditional and contemporary urban settlements.

**Abdalrahman Alashi** is an architect. He received his architectural education in 2013 from Misr International University (MIU), Cairo, and his M.Sc. degree in urban planning in January 2020 from Istanbul Technical University (ITU), Turkey. Alashi is currently enrolled in the Urban and Regional Planning Ph.D. Program at Istanbul Technical University (ITU). His master thesis was "Unmanned Aerial Vehicle based 3D City Modeling"; where a complete workflow of data collection, processing and analysis was developed, Yavuz Sinan neighborhood was the case study area. During his study, he contributed to a book chapter titled "Towards City Information Modeling", which was published in May 2019. Alashi has contributed in different scale architectural designs, he was enrolled mostly in the 3D architectural visualization. As both a Revit Architecture 2015 certified professional and Autocad Architecture 2015 certified professional. His main tasks were the development of 3D models, performing analysis, consulting the implementation and following up with clients.

**Diana Barrera** was born in Manizales, Colombia, in 1992. She graduated from Architecture at Pontificia Universidad Javeriana in Bogota. Her bachelor´s thesis project addressed rural development issues; Rural Reaction was a strategy for young people to be interested in living and developing in the countryside, avoiding large migrations to the city. She is an architect with interest in interdisciplinary and participatory projects with experience in competitions, urban, architectural and landscape projects. Her work has been focussed on rural development, urban redesign, and climate adaptation. Facing the Water was the first neighbourhood in Cartagena to be adapted to climate change; Binding Traces was a strategy to rethink and to redesign two main squares and the street that connects them in La Tebaida, Colombia, and Caravane is about designing a rural artisan village built with Guadua, the Colombian bamboo. The transversal objective in her work is to empower the communities in their territory and encourage local development to improve their quality of life. That is why the IUSD master programme is the key to continuing her path with a deeper focus on holistic urban and rural projects from a sustainable design perspective and strategies applicable in different scales -global, regional and local- and contexts -countryside and cities.

**Mohamed El Fayoumi**, PhD, is a specialist in architecture, urban design, landscape architecture, and their interaction with traffic, transportation, and road design. Since 1999, Fayoumi has taught and supervised a variety of multidisciplinary topics at Ain Shams University. Currently, he is an associate professor in the Department of Urban Design and Planning, Faculty of Engineering, Ain Shams University. He held several academic positions, including the assistant vice dean, unit head, and academic coordinator for the Landscape Architecture Program. He is currently the Planning and Urban Design Manager and a member of the Consultancy Unit's board of directors. El Fayoumi is the academic coordinator of the Ezbet funded project between Ain Shams University and Stuttgart University. He has also participated in and coordinated several international cooperation projects and workshops with universities in Germany, Italy, Sweden, Belgium, Hungary, and France in architecture, landscape design, urban design, transportation, and traffic.

**Nardine El-Bardisy** is a PhD student in the Urban Planning and Design Department at Ain Shams University and a project manager at MB Consultant working as a noise expert. She got hold of her Dual MSc degree from IUSD (integrated urbanism and sustainable design) from Stuttgart University, Germany and Ain Shams University, Egypt, in September 2019. She obtained LEED AP (BD+C) in August 2020. In November 2019, she published a paper from her master's work under "Social Interaction Dynamics Exploration: The Case Study of El Sherouk City. She received a BSc. in urban planning and design in 2016 from the Faculty of Engineering at Ain Shams University. She is enthusiastic about the idea of regional and strategic planning processes and sustainable development and architecture. During her undergraduate studies, she participated in diverse workshops in cooperation with Cardiff University in London, Antwerp University in Belgium, and Eotvos Lorand University in Hungary. Nardine has gained practical experience in various fields, including sustainable design, landscape design, regional planning, and environmental impact assessment. She worked on noise and vibration measurements at MB Consultant. She was also a trainee for multiple architecture consultants. In addition to her academic career as a teaching assistant at Future University in Egypt.

**Yara Elmaghrabi** is a Palestinian–Syrian Refugee who holds a Bachelor's degree in Architectural Engineering from Ain Shams University in Cairo, Egypt. She participated in designing a 300 Feddan Multi-Functional land in San Miguel, Mexico City, Mexico, and worked as an Intern at multiple international offices such as ECG and CLUSTER. After graduating in 2017, she worked as a Research Coordinator Assistant at IUSD where she had the chance to participate in and co-coordinate a handful of workshops and excursions. She engaged in projects addressing Climate Change in ElGouna, Resilient Communities in Nuweiba, and Informal Settlements in Helwan. To expand her professional experience, she worked at architectural consulting firms including Arch-Plan and IDG. She collaborated in designing residential, commercial, educational, and administrative projects in the New Administrative Capital and the New Alamein City respectively. Since 2016, she has also been freelancing as an Architect, a Ghostwriter, and a Photographer for multiple companies. Her aspiration to make a difference has led her to focus her research interests on Sustainable & Resilient Disaster Relief Housing, Participatory Design, and Usage of Public Spaces.

**Eman A. Elmeligy** is an M.Sc. student in the Urban Planning and Design Department at Ain Shams Univesity, where she also received her Bachelor's degree in 2020. She is currently working as a teaching assistant at the university she graduated from as a junior landscape architect at Oekoplan engineering consultancy firm. Eman is interested in the climate action movement and where it meets sustainable and resilient urban design and planning.

**Özhan Ertekin** is still working as a faculty member in the Department of City and Regional Planning at the Faculty of Architecture at Istanbul Technical University. His academic interests are mainly in regional planning, urban design, urban transformation, performance measurement and transportation planning. Some of his works, South East Anatolia Master Plan, Prizren Development Plan, Kazakhstan Mangistau Region Tourism Master Plan at the regional level, Zeytinburnu Urban Transformation Model, Istanbul Urban Transformation Strategic Plan and Master Plan at the urban regeneration level, many squares, street arrangement projects and competition achievements at the urban design level, Istanbul Main Artery Project and Istanbul Bicycle Roads Project in transportation issues.

**Riham Nady Faragllah** is an Associate Professor at the Department of Architectural Engineering in Pharos University in Alexandra (PUA). She has received her Ph.D degree from the Faculty of Fine Arts, Alexandria University (2013). She has an academic experience in teaching several courses such as architectural design, execution design, theories of architecture, environmental studies, technical systems in buildings, … etc. Also, she has published several scientific manuscripts in international journals and conference proceedings concerning urbanism, informal settlements, low energy systems and sustainability. She has participated in different architectural conferences internationally and locally. Her main research interests lie in the field of environmental studies and urban design.

**Ahmed Fayed** is a Teaching assistant at American University in Cairo, Egypt.

**Lina Fayed** is an Assistant Professor at the German University in Cairo (GIU). She obtained her bachelor's and master's degrees from the Faculty of Fine Arts at Helwan University. In 2019, she received her PhD from Ain Shams University in Cairo. Her research interests are in urban planning and design with a particular emphasis on regenerative city planning.

**Natalie Gulsrud** is an Associate Professor at the University of Copenhagen, Department of Geosciences and Natural Resource Management, Section for Landscape Architecture and Planning. She studies the governance of urban green infrastructure to advance sustainable and just pathways to climate resilience. This research engages social, ecological, and technological interactions in the "gray to green" infrastructure of cities from bicycle pathways to parks, trees, and community gardens. She publishes in scientific journals such as Landscape and Urban Planning, Environmental Research, and Urban Forestry and Urban Greening and is the co-author of the book "Street Fights in Copenhagen: Bicycle and Car Politics in a Green Mobility City" (Routledge, May 2019). She co-leads the Swedish government research council for sustainable development (FORMAS) "A sustainable spatial planning framework for engaging diverse actors and citizens in revitalising in-between spaces for social inclusion, biodiversity, and well-being" (VIVA-PLAN) in Denmark and Sweden. Other projects include WP leader in the NordForsk funded Smarter Greener Cities project and WP leader in the FORMAS funded Planning with Youth: a tool and framework for engaging meaningful and forward-looking participation of youth in shaping attractive and sustainable living environments. She also teaches the masters-level course Urban Forestry Urban Greening.

**Hamdy Hefnawy** is a young ambitious and dedicated Landscape Architect/Architect and Urban Design Scholar with professional experience in some urban design, architectural projects as well as international competitions. Hamdy is very interested in Architecture, Landscape urbanism and Landscape Architecture as well as is oriented to people and environmental research-based projects since he believes in making a difference through my promising designs.

**Heba Khalil** is a professor of sustainable urbanism at Department of Architecture, Faculty of Engineering, Cairo University and the senior coordinator of the Architectural Engineering and Technology AET program with 20 years of academic and professional experience. She teaches and conducts research in community development, informal areas, urban climate and integrated urban systems with several publications. She has built several international research partnerships focusing on urban metabolism and improving urban micro-climate in highly dense cities through material flow analysis, simulation and participatory action research. More recently, she has started a research project focusing on gender equity and cities and another on land governance. Professionally, she works as an architect, urban planner, community facilitator and housing expert with multi-scale agencies, authorities and clients both local and international. As an Alumna of SPURS, MIT, she continuously seeks partnerships to support a new planning and design practice for sustainable urbanism in her region. Khalil holds a PhD in architectural engineering from Cairo University and has been granted the Hubert H. Humphrey Fellowship in urban studies at MIT.

**Parisa Kloss** is a trained architect and urban planner with 18 years of experience in Middle East, Asia and Europe. She is Founder and CEO of Resilient Urban Planning and Development (RUPD) GbR in 2013 in Germany. Parisa was a post-doctoral fellow at Freie Universität Berlin, Institute für Meteorologie from 2011 until 2013. There, she focused on simulating urban heat island effects. She holds a PhD in urban climate with focus on ''urban heat island'' from The National University of Malaysia (UKM) and studied her Master and Bachelor in architecture in Iran. Her main research interests cover urban climate change planning; urban heat island effects; climate adaptive and intelligent cities, and urban resilience.

**Turgay Koramaz** has highlighted information and communication technologies, spatial analysis and modelling in urban planning, leading to publications on the subjects of cultural heritage management, spatial determinants of housing property prices, locational pattern of cultural venues, and internal migration patterns, in Turkey. Next to developing evidence-based decision-making processes in urban planning, methodological framing and information system, he has been involved in professional activities and consultancies in planning agencies and institutions. He is member of European Regional Science Association, ICOMOS, International Council on Monuments and Sites, and AESOP, Association of European Schools of Planning.

**Melvina Puspahati** is an Indonesian. She has graduated with her bachelor's degree in architecture from Sebelas Maret University, Solo. Currently, she is a master's student in the Integrated Urbanism Sustainable Design Double Degree program at the University of Stuttgart, Germany, and Ain Shams University, Egypt. Her master's study is fully supported by the DAAD scholarship. Melvina has six-year of experience as an urban designer in a national and international company. She was responsible to manage and distribute the workload to her team project.

**Alaa N. Rashed** is an MSc student in the Urban Planning and Design Department at Ain Shams University, Cairo, Egypt. She obtained her bachelor's degree in architecture engineering from Ain Shams University in 2021. Alaa works as an architect for a construction company named Concrete Plus as a part of a team currently working on residential projects in the New Egyptian Capital City. She is very interested in developing her local community and comes up with new ways to help people have a better life.

**Shereen Wael**, BSc, is an architect and a teaching assistant at the Department of Urban Planning and Design, Faculty of Engineering, Ain Shams University (ASU), Cairo (tenure track). She got her bachelor's degree in 2018 Urban Planning and Design specialisation from ASU. Her interest is in urban design, with a special focus on the underground stations and the various effects of urban morphological configuration. Professionally, she worked in the ArchPlan firm as an architect and researcher. She has participated in some technical projects.

**Dilek Yildiz Ozkan** graduated from Istanbul Technical University (ITU) Faculty of Architecture, Department of Architecture in 1992. She received her master's degree from ITU Institute of Science and Technology, Architectural Design Program in 1996. She has been a visiting researcher at "North Carolina State University, College of Design" in 2001. She completed her Ph.D. with the thesis titled "A Multi-Factor Analysis Model to Determine the Use Value of Enclosed Outdoor Spaces" at ITU Institute of Science and Technology, Architectural Design Program in 2004. She was appointed to the Faculty of Architecture, Department of Architecture as a research assistant in 1995 and as an assistant professor in 2005. She received the title of associate professor in 2017 and is still in the same position at ITU. She continues to work as a lecturer/academician at the Faculty of Architecture, Department of Architecture. Her research interests include architectural design theory, research methods, environmental-behavior studies, neuro-architecture, housing design and its problems, sustainability, urban space, urban design, urban architecture. She has participated in many national and international scientific conferences, researches and workshops related to her fields of study.

# Index

www.ingramcontent.com/pod-product-compliance
Lightning Source LLC
LaVergne TN
LVHW081328300926
849161LV00008B/1911
*9781668424629*